HUMAN ANATOMY

HUMAN ANATOMY

José M. Parramón

Watson·Guptill Publications/New York

Copyright © 1990 by Parramón Ediciones, S.A.

First published in 1991 in the United States by Watson-Guptill
Publications, a division of BPI Communications, Inc.,
1515 Broadway, New York, NY 10036.

Library of Congress Cataloging-in-Publication Data

Parramón, José María.
   [Cómo dibujar la anatomía del cuerpo humano. English]
   Human anatomy / José M. Parramón.
      p.   cm.—(Watson-Guptill artist's library)
   Translation of: Cómo dibujar la anatomía del cuerpo humano.
   ISBN: 0-8230-2499-7
     1. Anatomy, Artistic. 2. Drawing—Technique. I. Title. II. Series.
   NC760.P3413 1991
   743'.49—dc20   90-49246
                                             CIP

Distributed in the United Kingdom by Phaidon Press, Ltd.

Manufactured in Spain
Legal Deposit: B-2.091-91

1 2 3 4 5 6 7 8 9 / 95 94 93 92 91

Contents

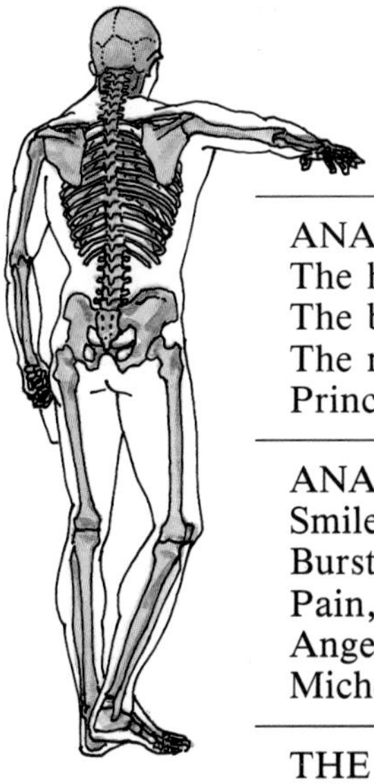

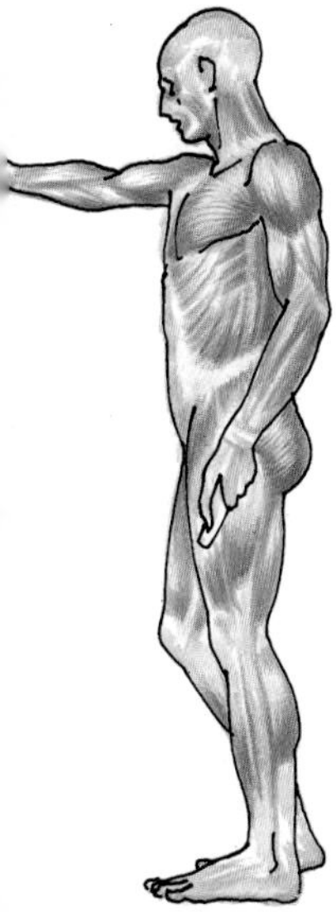

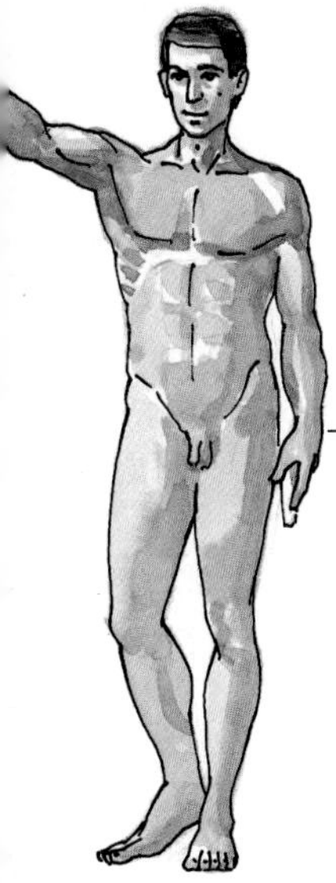

Fig. 1. (Previous page). Copy of Michelangelo's fresco *The Creation*, Sistine Chapel.

2

Fig. 2. Bartolomeo Passa-
rotti (1529-1592), *Les-
sons in Anatomy*, Bor-
ghese Gallery, Rome.
During the Italian Renaissance corpses were dis-
sected as a scientific way
of studying anatomy.
Although dissection was
illegal, some great
masters of the time did it,
such as Michelangelo
Buonarroti, whom we
see drawing from life, sitting on the far right end
of the picture.

Introduction

"Your stomach may perhaps object to your spending the night in the company of corpses, bones and human flesh, all opened up and in pieces..."
The words are from the *Treatise on Painting* written in the fifteenth century by that genius Leonardo da Vinci, painter, poet, musician, architect, inventor, and also the most courteous and elegant man in Florence during the Renaissance. According to Renaissance artist and historian Giorgio Vasari, Leonardo wrote that "after having minutely examined some corpses and suffering terrible nausea and sickness, he wanted to discover the secrets which govern the parts of the human body."
Also in Florence at the same period, Michelangelo Buonarroti was charged with profaning corpses "for having dared, in the pursuit of knowledge, to cut open from top to bottom the dead body of a member of the Corsini family, another of whom stirred up a tremendous fuss and prosecuted him before Piero Sonderini, who was Chief Magistrate at the time."
Dissecting bodies, even for scientific purposes, was prohibited and punished by imprisonment. So those great men, who even then were considered geniuses, had to bribe servants in the hospitals in order to learn the rules of anatomy for the artist. Imagine them working at night by candlelight in the morgue of the Santo Spirito Hospital in Florence, hushed and alert, knowing they risked being put in jail.
A working knowledge of human anatomy is essential for artists; fortunately, today there are easier ways to learn it.

José M. Parramón

Generally speaking, anatomy for the artist can be divided into two main parts: knowledge about the *bone structure*, including joints; and *muscular structure*, which also includes morphology, or the external appearance of the human body. The muscle structure is usually the most appealing for the artist, since it determines the shape and look of the body. In the first chapter we will deal with the anatomy of the human head, its bone structure and the many small muscles that cover it. Each of these muscles has a special purpose, and together they control all the facial movements by which we express our emotions.

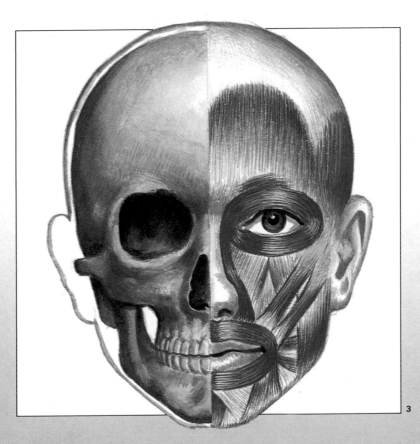

3

ANATOMY
—OF THE—
HUMAN HEAD

The human head: its bone structure

Figs. 4 and 5. Except for the mandible, the bones in the cranium lack movement. Fig. 5 shows how the bones are linked to each other, covering the cranial cavity.

You will probably find this the easiest way to use this book: first read and study the explanations in the text paying careful attention to the illustrations; then, examine your own body to see how it fits in with what you have been studying. Use your fingers to try to locate the bones (in this chapter, it would be the facial bones), checking their shape and size.

The bones of the skull

Basically the skull consists of the following groups of bones: frontal, occipital, parietal, temporal, ethmoid and sphenoid. Only one is of any great interest to the artist: the *temporal* bone.

In this front view of a skull note the location of each of the temporal bones in relation to that long bone known as the *zygomatic arch* (the cheekbone); notice how this cheekbone projects outward considerably from the plane of the temporal bone and then you will appreciate that in some faces—lean, thin, bony, or old faces mainly—this slight depression or concavity of the temporal bone will produce a characteristic difference in level between the plane of the ear and the end of the jawbone.

Also on the temporal near the orbital cavity (eye socket) study that unexpected hollow that in anatomy is called the *temporal fossa*. On the edge of it, beside and above the orbital cavity, there is a visible ridge, called the *arista*, which is indicated in Figs. 4 and 5 by the letter A. Both the temporal fossa and the arista will be visible in bony or delicate faces, where absence of fat makes the bone structure more prominent.

And that is all we need mention about the cranial bones. Let's go on to the main bones of the face.

4

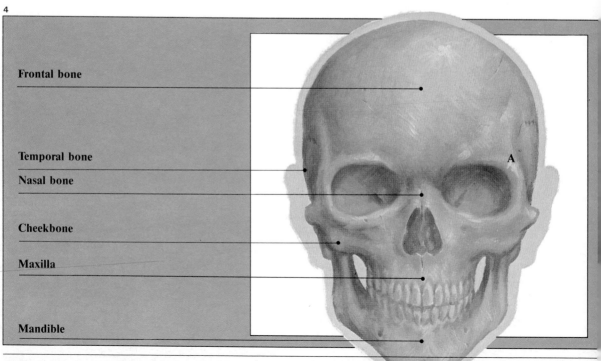

Frontal bone

Temporal bone

Nasal bone

Cheekbone

Maxilla

Mandible

A

The bones of the face

The malar or cheekbones

Notice the exact position of this important pair of bones. You will see that the *malar* (cheekbone) lies right below the orbital cavity, projecting above the jawbone (*mandible*); notice, too, that hollow, empty area below the cheekbone. Study the shape of the malar bone as it extends backward, forming that long bone called the zygomatic arch. Now imagine the entire area covered with skin and flesh; visualize the shape of the lower face. Notice that the hollow we call "the cheek" can be deep or slight, depending on how much fat covers it. In a thin face the entire shape and size of the malar will be visible, producing a hollow, sunken cheek.

Study these important features in your own face in front of a mirror; try to visualize the bone under the skin, imagining its exact position and size.

The nasal bones

These two small identical bones meet at the center of the face, forming what is known as the bridge of the nose. The lower part of the nose below the bridge is made up of cartilage, a kind of hard, elastic tissue that holds the tip of the nose firm while still allowing a certain amount of movement. The important point for artists to remember is that the lower ends of the nasal bones (B) can be precisely located when you draw a nose, and they determine the general shape of the nose: straight, upturned, aquiline, and so forth.

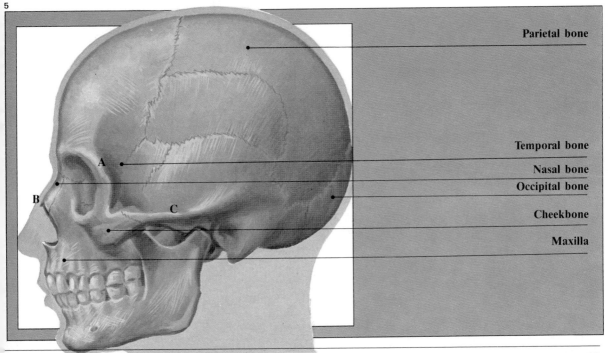

5

Parietal bone

Temporal bone

Nasal bone

Occipital bone

Cheekbone

Maxilla

The maxilla and the mandible

The maxilla

This bone holds the teeth along its lower edge in the shape of a horseshoe. It is worth remembering this horseshoe shape. Also remember that the maxilla forms part of the cranium; *it cannot move independently*.

The mandible or jawbone

Its importance for the artist, the need for careful examination, and the difficulties caused by that examination are all due to one basic fact: *The jawbone is mobile*. It is the only bone in the head that can move; it is lowered and raised by the action of certain muscles when the mouth is opened or closed to laugh, shout, groan, or cry.

The mechanism controlling this movement is very simple: at the rear end of the jawbone (marked C in Fig. 5 on the previous page, in the profile skull) where it joins the temporal bone, there is a small knob (in anatomy it's called a *condyle*) set perpendicular to the primary line of the jawbone. This knob fits into a corresponding depression (point C), creating a kind of hinge to support the jawbone, allowing it to open and close. Study the simplified version on this mechanism in Fig. 6. The drawings marked A show two small semicircular shapes that represent the maxillary cavity. In the next drawing (B) I have fitted a pair of brackets, one straight, the other forming a right angle, which are joined at a' by a spindle representing the condyle.

Now imagine this small device opening and closing; for each movement study the position adopted by the maxilla and the mandible (C). Notice especially the form taken by the edges representing the tip of the teeth (b' and c'), the special way they separate and come together when the mouth is opened and closed, and how this function is connected with the hinge (a').

Finally, if we apply the simplified form of this movement to the skull and then to the external shape of the mouth (Fig. 7), we could say that we have completed our study.

But we haven't! Take a pencil and paper and start copying the illustrations on page 13; you'd be wise to spend a little time on this. First draw the mechanical device representing the movement of both maxillaries; show them closed, half-open, fully open, seen from above, below, and so on. Then, copy the sketches of the skull and mouth. Finally, imagine other positions from other angles, drawing them from memory and from life, using your own face in the mirror as the model.

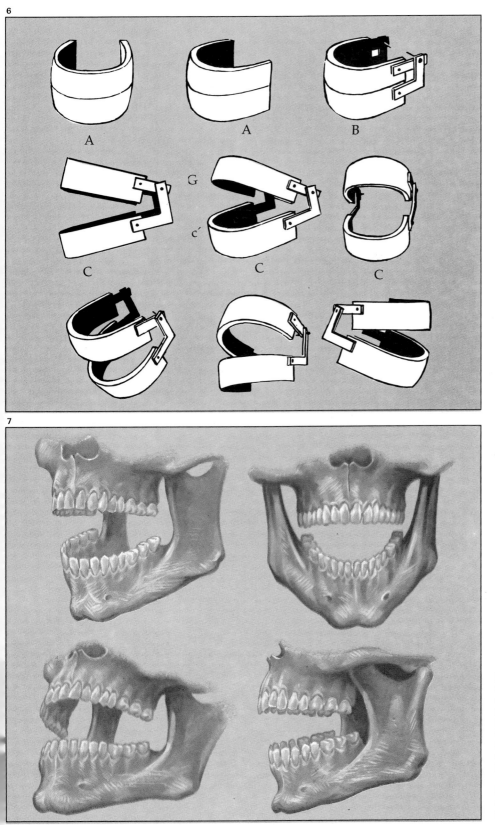

Fig. 6. This mechanical model can reproduce almost every movement of the jaw. Study carefully every point of view and the several positions it adopts. Then you'll be able to draw the maxillaries quite well.

Fig. 7. These drawings acquaint you with the movement of the jaw.

Principal muscles of the human head

The head muscles are divided into two categories: the masticatory muscles, including the *temporalis* and the *masseter*; and the mimic muscles that control expression—the *frontal, supercilius orbicularis* of the eyes and the lips, *levator* of the upper lip, *buccinator*, the *triangularis* muscles of the lips, the *depressor* of the lower lip, the *minor* and *major zygomatics* and the *risorius* (Fig. 8).

As their name suggests, the masticatory muscles mainly produce chewing action. The mimic muscle activate facial movements, wrinkles caused by worry, pain, grief, or laughter. Some of these muscles lie just under the skin or are interlaced with other muscles—or, to put it another way, they are not linked to any specific bone.

We are going to study each one in turn with some thoroughness since, to a large extent, your ability to portray facial expression depends on understanding these muscles and how they work.

All the muscles named above occur in pairs, such as the two temporales, masseters, and so on. For simplicity, however I will describe only one of each pair. Please bear in mind that I am in fact referring to the pair.

8

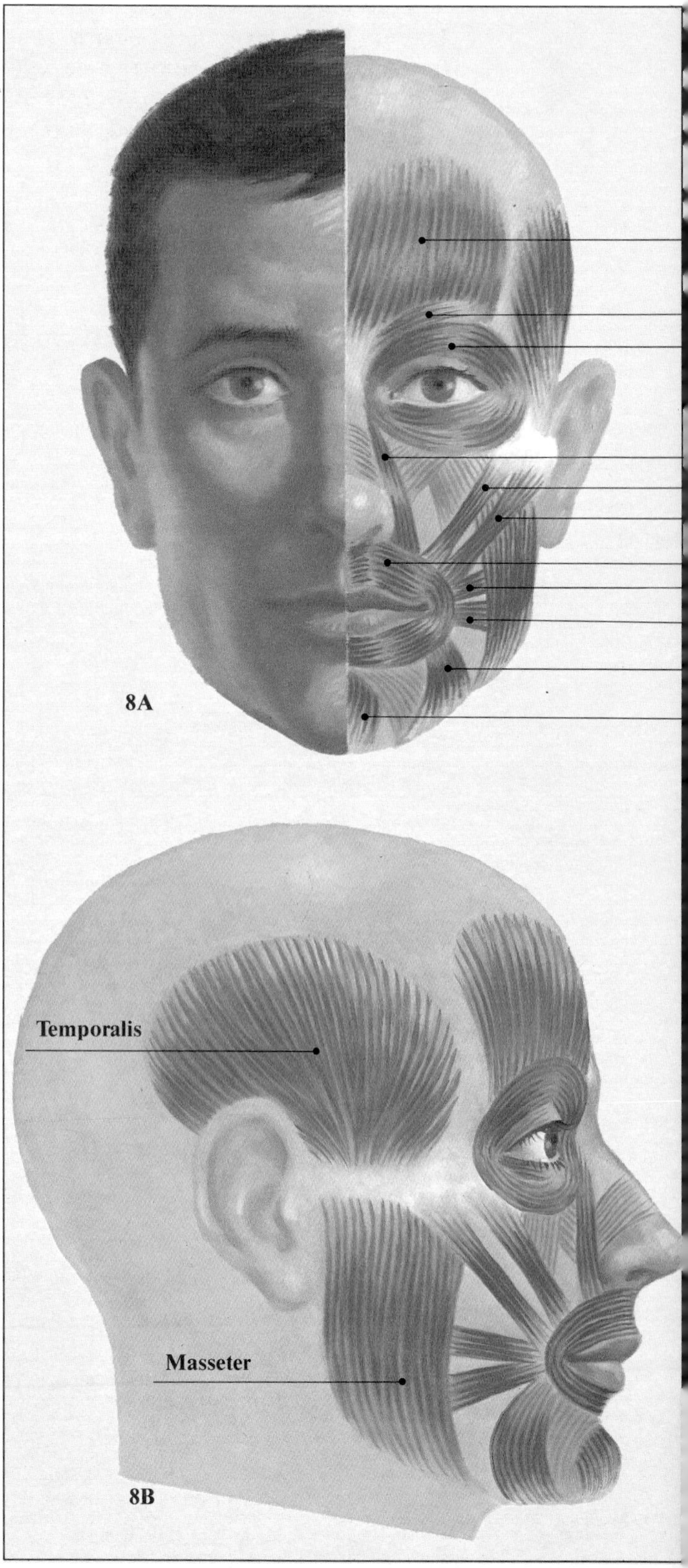

8A

Temporalis

Masseter

8B

Fig. 8. The human head, especially the face, has the most muscles of any area in the body. Through these muscles we perform many functions: breathing, smelling, chewing, and expressing emotions.

Frontalis

Supercilius

Orbicularis of the eyes

Levator of the upper lip

Zygomatic minor

Zygomatic major

Orbicularis of the lips

Buccinator

Risorius

Triangularis muscles of the lips

Levator menti

Temporalis

If you close your mouth, clench your teeth, and then touch your temple with your fingertips, you will see there the indication of this important muscle (Fig. 9). (Try this now in front of the mirror.) Think about the occasions when you have to close your jaws with some force ... when you are exasperated, offended, or angry; in other words, whenever you express violence. This muscle is also used when we chew, but artists mostly think of this as the muscle through which we produce an expression of violence, anger, or hate. The temporalis covers the temporal bone. It is shaped like a fan whose base is joined with the mandible after passing under the cheekbone. When contracted, it closes the mouth by pulling the mandible upward until it meets the upper. It has astonishing power.

Masseter

This muscle is found in the lower angle of the jaw and helps to close the mouth and clench the teeth (Fig. 10). We use it to chew by contracting and relaxing it simultaneously with the temporalis. The top is inserted in the extension of the malar bone and the bottom lies in the lower edge of the jawbone. This muscle, too, is noticeable in violent expressions when the mouth is closed.

Figs. 9 and 10. The temporalis muscle (Fig. 9) and the masseter (Fig. 10) are directly involved in chewing. If you clench your teeth you can notice that they stand out on the temple and the lower jaw. That is why they're so important in expressions of anger and violence.

9

10

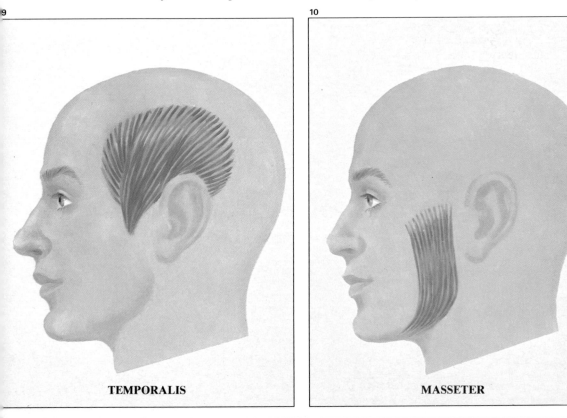

TEMPORALIS　　　　　　　　　　**MASSETER**

Principal muscles of the human head

Frontalis

The remaining eleven muscles are all mimic muscles, which produce expression. The *frontalis* is the muscle of surprise, horror, and terror... and also joy. This thick, flat muscle covers the forehead and, when it contracts, it produces the characteristic horizontal wrinkles in the forehead. It also raises the eyebrows and the upper eyelid when we express horror or surprise.

Supercilius

This muscle comes into play when people "knit their brows"—when they are worried, puzzled, thinking hard about something, in pain, crying, and so on (Fig. 12).

The *supercilius* is a small bunch of muscles lying exactly under the eyebrows. When contracted, they pull the brows toward the center, producing vertical wrinkles between the brows.

Look in the mirror and frown. Do you see them?

Orbicularis of the eyes

In addition to opening and closing the eyelids—which we do constantly and instinctively—this muscle is involved in every expression of the human face. It forms a ring around the eyes. When it contracts, mainly in expressions of unhappiness, it half-closes the eyes and accentuates the wrinkles at the outer corners of the eye that we call "crow's feet." It also produces similar but larger wrinkles in expressions of happiness, when we burst into laughter.

Buccinator

We use this when we blow. When we fill our mouth with air, this muscle relaxes as if it was swollen and then, when it contracts, it reduces the space in the mouth, which in turn compresses and expels the air. We use it to whistle, play a wind instrument, or simply to cool our soup. The buccinator is located in the cheeks, starting above the gums and ending at the corners of the the mouth.

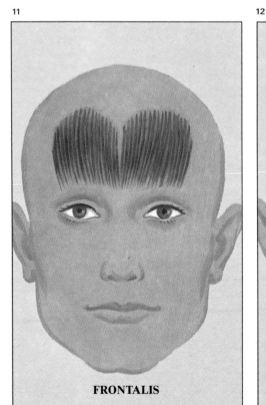

11

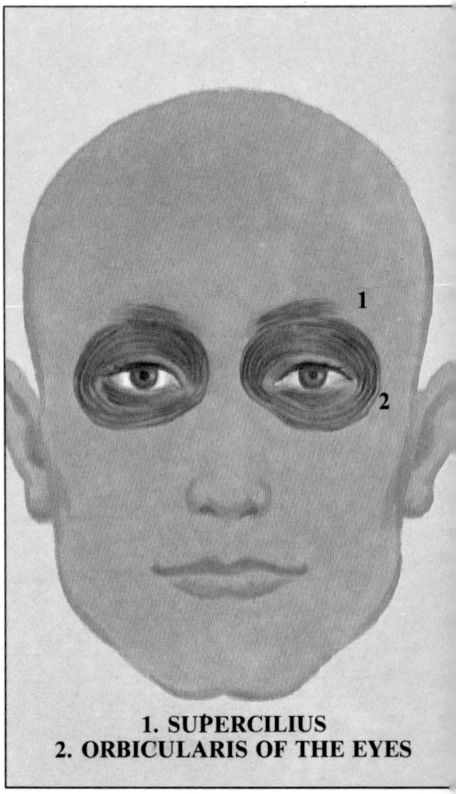

Fig. 11. The frontalis muscle raises the eyebrows and produces wrinkles on the forehead. It is the muscle of horror, surprise, and laughter.

Fig. 12. The supercilius muscle (1) controls the knitting of the eyebrows, pulling them toward the center. The orbicularis of the eyes (2) allows the eyelids to move and is involved in every expression around the eyes.

FRONTALIS

12

1. SUPERCILIUS
2. ORBICULARIS OF THE EYES

Orbiculars of the lips

Without this you couldn't kiss or drink through a straw, for this is the muscle that allows us to purse the lips. But the more common function of the orbicularis is to open and close the lips. It can do this without moving the maxillary, that is, while the teeth remain closed. It surrounds the mouth in the shape of a ring and, like the orbicularis of the eyes, is involved in almost every expression.

Levator of the upper lip

This is called the muscle of pain because it plays a leading role in expressions of hurt, grief, or weeping. By itself it can produce a perfect expression of disdain, scorn, and, with very little extra help, repugnance. When contracted, it raises the upper lip and simultaneously dilates and lifts the nostrils. (These movements also involve other muscles, which we need not mention here.) Its upper part is attached to the nasal bones; at the bot-

tom it splits into two bunches, one of which connects to the nostrils and the other to the upper lip.

Triangularis muscles of the lips (Fig. 16, below)

These help the levator in expressions of pain, grief, and weeping. They are also involved in many other expressions such as disgust, repugnance, fear, and anger. They lie on each side of the mouth in the form of a triangle whose peak starts almost at the outer corner of the lips. When contracted, they lower that corner point.

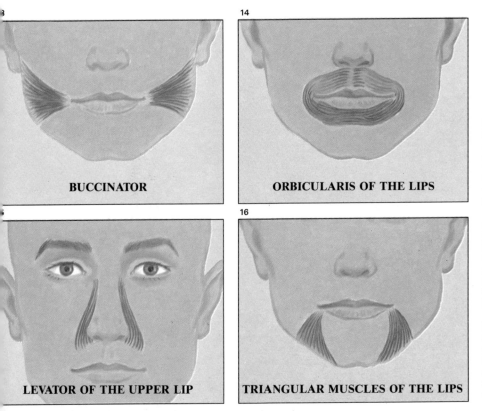

BUCCINATOR

ORBICULARIS OF THE LIPS

LEVATOR OF THE UPPER LIP

TRIANGULAR MUSCLES OF THE LIPS

Figs. 13 and 14. The buccinator (Fig. 13) allows us to contract and enlarge the space in the mouth. The orbicularis of the lips (Fig. 14) is involved in any movement to open and close the mouth.

Fig. 15. Apart from the lifting function implied by its name, the levator of the upper lip dilates the nostrils and forms the folds that extend from the nostrils toward the lip.

Fig. 16. The triangularis muscles of the lips push the lips downward toward their outer corners, helping in expressions of disgust or repugnance.

Principal muscles of the human head

Levator menti

This is the muscle of anger, fury, and all aggressive expressions. It acts in combination with the other muscles surrounding the lips, raising the chin and the lower lip and at the same time, pushing the bottom lip outward, producing an expression of furious aggression. This small muscle lies in the chin with its top end in the mandible and its base at the point of the chin.

Zygomatic minor

This is the cry-baby of the family. When it contracts, it raises and forces forward the middle of the upper lip, changing the curve to express pain or grief. It consists of a small bunch of muscles across the cheek, extending from the upper lip to the cheekbone.

Zygomatic major

Now smile—you are using this kindly, joyful muscle. It plays a vital part in smiling and laughing. It stretches diagonally from the cheekbone down to the corner of the mouth. When it contracts, it widens the mouth and pulls the corners upward, puffs out the cheeks and, at the same time, accentuates the wrinkles or crow's feet below and beside the eyes.

Risorius

This is also a laughter muscle. It helps the zygomaticus major by pulling back the corners of the mouth. It is very small and so we have left it until last. It starts in the cheeks and stretches horizontally to end near the corners of the mouth.

Fig. 17. The levator menti contracts the chin and raises it, at the same time raising the lower lip and pushing it outward. It produces expressions of intense aggression.

Fig. 18. When the zygomatic minor contracts, it raises the middle of the upper lip, expressing pain or grief.

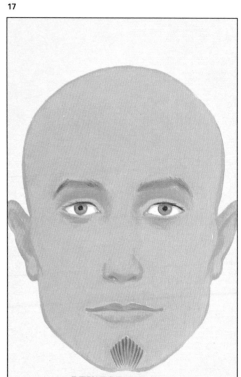

17

LEVATOR MENTI

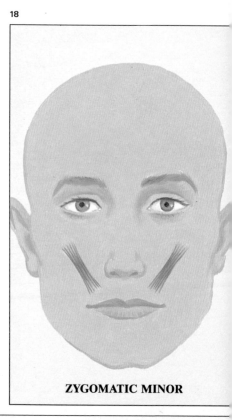

18

ZYGOMATIC MINOR

And that's all. These short lessons should be enough to give you a good idea of the facial muscles and their main and subsidiary functions. After you become familiar with the names, positions, shapes, and sizes of all these muscles, you will be ready to move on to the next section, where we shall study various expressions of the human face and work out the muscles that relax or contract to produce each one.

Obviously, a muscle by itself is hardly ever capable of expressing an emotion perfectly. For instance, when the risorius operates without the others, it produces a smile that looks forced—this is a special expression in itself. To produce a sincere laugh or smile, several muscles must play their part at the same time: the orbicularis of the lips, the two zygomatic muscles, the risorius, and the frontalis, which raises the eyebrows and wrinkles the forehead to express cheerfulness. It is this combination of the various muscles in the creation of a specific expression that we will now study.

Figs. 19 and 20. The zygomatic major (Fig. 19) helps the risorius (Fig. 20) in joyful expressions: The former pulls the corners of the mouth upward and outward when we laugh; the latter stretches the corners of the mouth horizontally when we smile.

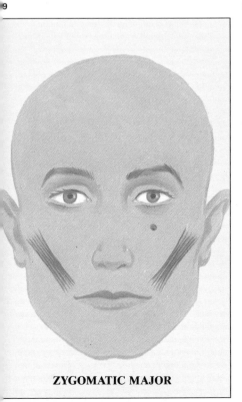

ZYGOMATIC MAJOR

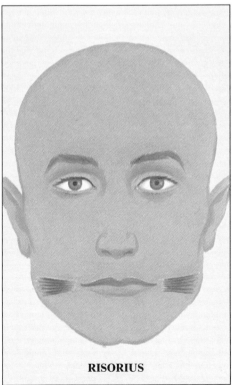

RISORIUS

I advise you to warn your family (or perhaps lock the door of your room) before you start on this section. We are going to study ourselves in the mirror, learning which muscles contract when we produce the various expressions. You will have to watch yourself frowning, suffering, laughing, doubting. My wife is accustomed by now to seeing me in front of the mirror, drawing my own face with a stupid or furtive look, but the other day when I was drawing one of the faces illustrating this lesson, she came into my studio and saw me with untidy hair, holding my head at a peculiar angle and scowling ferociously. "You really should lock the door," she said. "If the children were to come in and see you like that it would be the end of them."

Another thing: It is not enough just to watch your expressions in the mirror—you must also sketch them. Draw your own laugh, your own look of grief, your forehead, your eyebrows and eyes while your face expresses worry, delight, surprise. Apart from the fact that you will enjoy yourself, you will learn much more.

ANATOMY
—AND—
EXPRESSION

Smiles and laughter

22

Smiles

The only muscles that visibly contract when we smile are the zygomatic major and the risorius. These two pull the lips apart slightly by raising the corners of the mouth and moving them sideways, producing a smooth upward curve at the junction of the lips. Be sure to notice that the skin at the sides of the mouth puffs slightly, causing a small crease from the nostrils to the lips. On the rest of the face a smile produces a slight change of position in almost every facial muscle. The lower section of the orbicularis of the eyes, for instance, contracts a little, lifting the skin of the lower lid and forming a delicate wrinkle under the eyes. The supercilius and frontalis muscle may raise the eyebrows a bit and produce shallow furrows in the forehead, or they may relax, causing the special wrinkles expressing attention or worry to disappear.

Laughter

Watch carefully what happens to our face when we laugh uninhibitedly. Laughter is often portrayed, especially in illustrations; laughing faces are used in advertisements and brochures.

Study the illustrations (Figs. 22 and 23) and notice the changes produced in the shape of the mouth by the contraction of the zygomatic major and the risorius, especially the former. See how this contraction causes the characteristic wrinkles to appear at both sides of the mouth. Observe, too, how this contraction puffs up the cheeks by tightening and folding the skin to produce wrinkles under the eyes. Notice that the eyes are partly closed, bright and smiling, in harmony with the laughing expression, so that the crow's feet at the sides are accentuated. Study the movement of the forehead and eyebrows and you will realize that the various kinds of laughter—a soft laugh, an active laugh, or a burst of laughter—affect the muscles in the eyebrows and forehead differently.

Fig. 21. (Previous page). Copy of a drawing by Leonardo da Vinci.

The most important aspects to study are the shape of the mouth, the lips and the corners of the mouth, the position and shape of the lines on either side, and the appearance of the teeth.

Figs. 22 and 23. Study these drawings carefully; observe the function of the zygomatic major and the risorius, widening the mouth and pulling the corners sideways and upward. Note the folds and wrinkles around the eyes and on the forehead, the raising of the brows.

23

Smiles and laughter

Let us try to analyze a standard laugh, the sort that occurs on every type of face. We must observe the precise position and action of the zygomatic major, which plays the leading role. Imagine that you are pulling it upward and sideways, stretching the mouth, while the maxillary bone and the other muscles are opening it. This is what you must remember:

1. Because of the action of the zygomatic major and the risorius when we laugh, the corners of the mouth usually form a small curve instead of terminating at an angle.
2. The upper lip is raised and stretched by these muscles and so moves upwards to adopt a horizontal position. But the lower lip seems curved.
3. Because both lips have been raised, only top row of teeth is visible.
4. The "laughter lines" at the sides of the mouth are caused mainly by the contraction of the zygomatic major. The top of these wrinkles coincides with the place where that muscle is attached.

Study the shape of these wrinkles and try to visualize the muscle underneath them. Getting their position and shape accurately is essential if the laughing expression is to seem natural. Notice that these "laugh wrinkles" start some distance away from the corners of the mouth and that their outermost point lies just above the corners of the mouth. When you laugh your lips become thinner and stretch so that the skin around the lips is taut. Finally, this important detail: the way of drawing the teeth. Beginners usually draw them exactly as they think they see them, with each tooth perfectly shaped and carefully placed on the curved gum, with the lines between them clearly shown.

Fig. 24. In general terms; laughter is illustrated by leaving the upper lip in a horizontal position while the lower lip curves, making the top row of teeth visible. Remember the importance of having the corners of the mouth forming a curve instead of terminating at an angle.

Fig. 25. Don't indicate the separations between the teeth; it's unrealistic and displeasing.

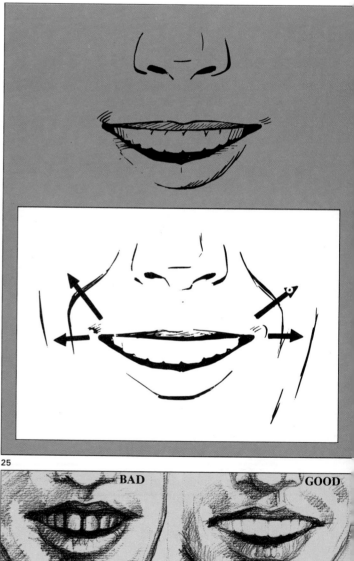

24

25 BAD GOOD

No matter how beautifully drawn, these look like the teeth of a very sick man; they attract our attention and spoil the effect. Avoid this hazard by remembering this rule:

Teeth should be drawn simply and sketchily, without the lines indicating the separations between them. Generally speaking, they should appear as a single white strip.

As you can see, the trick is to draw a simple white strip *with the shape of the teeth marked just along its bottom edge*. You can also draw the shape of the back teeth at both sides using very tiny pencil strokes.

As the last point in this study of laughter, notice the shape of the mouth—the lips, teeth, wrinkles, and so on—when seen from different angles: front view, three-quarter view, in profile, and so on (Fig. 26).

Fig. 26. To better understand the mechanism of laughter try to reproduce these drawings, showing the same expression from different points of view. Don't forget that life comes to the drawing when both the lips and the folds caused by the zygomatic are well represented.

Bursts of laughter, grief, crying

Bursts of laughter

These involve all the muscles used in the ordinary laugh, including those that raise the brows and wrinkle the forehead; in a sharp burst of laughter these muscles contract with more force. The head is also bent backward and the mouth opened wider, creating more lines than an ordinary laugh.

Notice the teeth during a burst of laughter; both the top and bottom rows are visible, particularly the top. This causes considerable problems of construction and perspective, which can be solved by examining the mechanism governing the movement of the jaw (Fig. 24, on page 24).

Grief

From the anatomical point of view, grief entails a kind of muscular abandon, a general relaxation in which only the muscles in the eyebrows and those that raise the upper lip remain tense. Imagine the laughter muscles in a completely relaxed state: the corners of the mouth drop and the muscles closing the jaw also ease, so that the entire cheek appears to be stretched. The eyelids also appear to be more closed than usual, producing that dead look that is so characteristic of grief. The supercilius in the eyebrows, however, slightly contracts, pulling the eyebrows inward and slightly puckering the forehead.

Crying

When a person is crying, the upper lip is raised a little and the triangularis muscles of the lips draw the corners of the mouth downward. The point of the chin may tremble slightly, because of the spasmodic contraction of the levator menti, which pushes the lower lip upward and a little outward. The nostrils dilate in conjunction with the agitated breathing caused by the sobs. In the upper part of the face the supercilius contracts, raising the eyebrows and bringing them toward the center of the face, producing small vertical lines on the forehead.

27

Fig. 27. In an outburst of laughter, all the changes caused in the face by laughter are emphasized: the raised eyebrows, the area of the eyes, and the open mouth showing both rows of teeth.

28

Fig. 28 and 29. Grief causes the facial muscles to relax (except for eyebrows and upper lip); crying causes them to contract spasmodically.

29

Pain, worry, self-satisfaction, envy

Pain
In physical pain the jaw is slightly open or firmly closed. In both cases the lips separate to show both rows of teeth. The lips are open in an arc (like the classical masks symbolizing tragedy); all the muscles around the mouth are involved in creating this arc shape. The contraction of the grief muscle, the zygomatic minor, becomes particularly evident, pulling the upper lip upward and outward. This strong contraction produces furrows on both sides of the mouth and, in conjunction with the orbicularis, causes marked wrinkles under and at the sides of the eyes. Notice the shape of the eyes when they express pain; they are half-closed, but in a nervous manner, bringing out a large number of lines above, below, and beside them. Notice, too, the important part played by the supercilius, which puckers the entire forehead so that these wrinkles are interlaced with those produced by the frontalis muscle. Finally, remember that when people feel pain, they instinctively tend to lift their head and tilt it backward.

Worry
Someone who is worried usually shows it by wrinkling the forehead and lifting the eyebrows, while at the same time bringing them closer together. The mouth remains firm and horizontal (in some cases this can emphasize the masseter) and the jaw is closed. When the jaw is very firmly closed, the face expresses vexation.

Self-satisfaction
When people are feeling smug, scornful, or haughty, we say they have their "head in the air." This is reproduced graphically by raising the eyebrows to varying extents, lowering the upper lids, and slightly tilting the head up and back. The disdainful set of the mouth is produced mainly by the contraction of the triangularis muscles pulling the corners down while the levator of the upper lip and the levator menti working together lift the middle of the lips.

30

Envy
For envy, suspicion, disquiet, or distrust, the orbicularis of the lips tightens, pursing the lips and accentuating their profile. The nostrils remain taut and wider than usual and the eyelids fall slightly to give the eyes a rather askance look.

Fig. 30. The expression of pain is associated with the downward arc of the lips and with wrinkles on the forehead, caused by the contraction of the frontalis and supercilius.

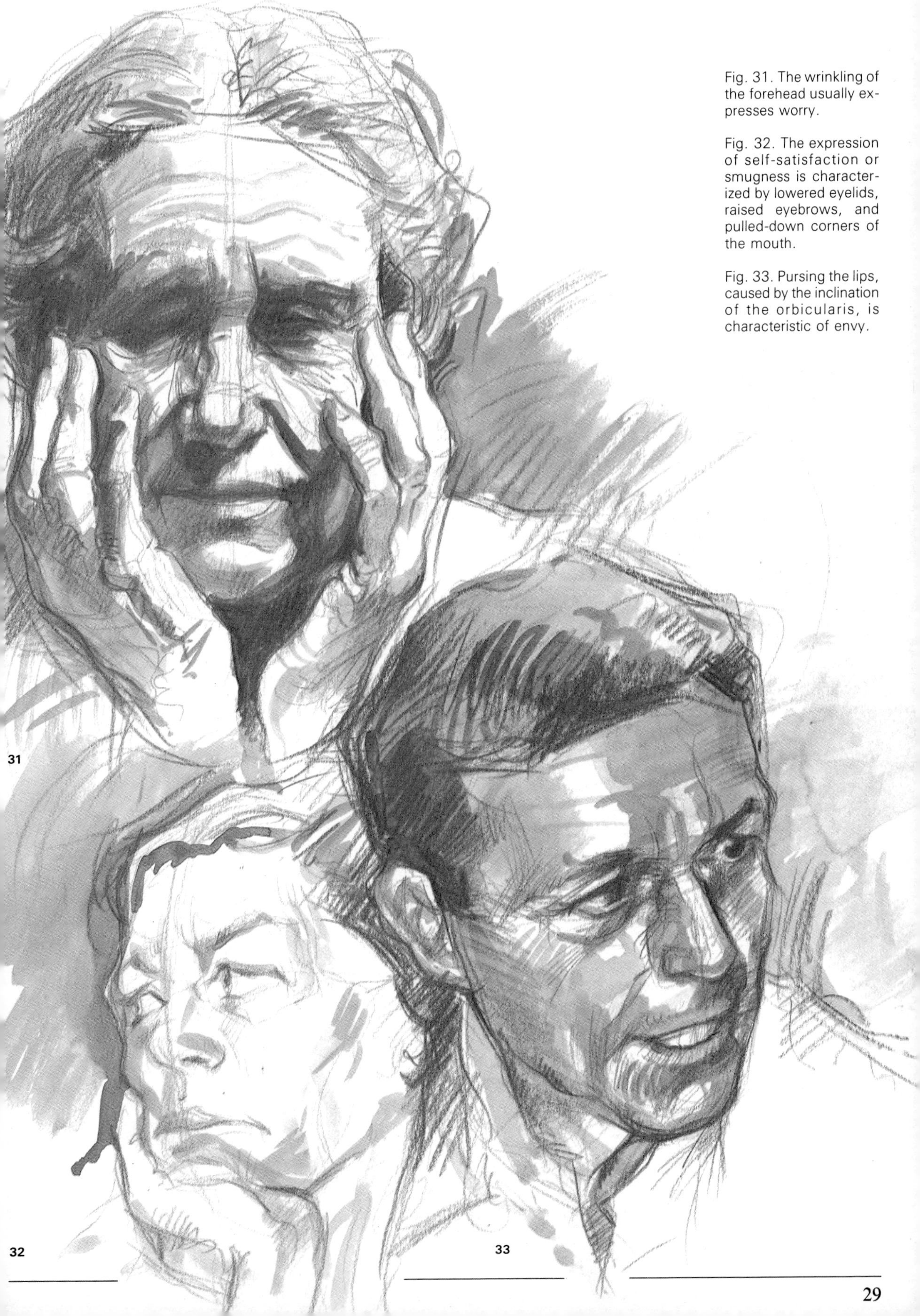

Fig. 31. The wrinkling of the forehead usually expresses worry.

Fig. 32. The expression of self-satisfaction or smugness is characterized by lowered eyelids, raised eyebrows, and pulled-down corners of the mouth.

Fig. 33. Pursing the lips, caused by the inclination of the orbicularis, is characteristic of envy.

31

32

33

Anger, exasperation, hate, fear, terror

Anger

In an angry or violent expression, the jawbone juts forward, extending the lower section of the maxillary cavity. Any speaking is liable to be more like shouting, with the mouth open wide. At the same time, the triangularis muscles pull the corners of the mouth down sharply and the zygomatic minor raises the upper lip.

Notice the deep furrows on both sides of the mouth, causing wrinkles below the eyes. The brows contract, creating furrows between them, and are pulled downward with a violent movement that wrinkles the forehead. The eyes are partly hidden by the contracted eyebrows, but the irises seem to blaze with fury.

Exasperation

Exasperation is a sudden, intense feeling of irritation. Most people inhale sharply and then let the breath out in a loud sigh. The eyes are usually open wide, the eyebrows lifted, and the mouth is set tight or slightly open, with the muscles around the mouth contracted tightly.

Hate

When you try to draw the look of hatred, you must remember that the jawbone is tightly closed, emphasizing the masseters; the mouth is also closed to form a horizontal line, with thin lips, and the corners slightly pulled down. The eyes, which are slightly more open than usual, show a blazing look, though this is muted by the rigid, horizontal position of the brows, which should be as low as possible with slight furrows between them.

Fear

Fear means that the eyes are open wider than usual and the eyebrows are raised; the mouth seems about to exclaim "Oh!" The eyes and brows are easy to draw; you have to arch the brows upward with slight lines between them, indicating the worry that accompanies fear. Portray the lids more open than usual, revealing the whites of the eyes with the iris and pupils more conspicuous than usual. The mouth is held open by the orbicularis and the two triangular muscles, which tend to round it slightly and lower the corners.

Terror

Terror or extreme fear is nearly always accompanied by a scream. All the mouth muscles come into play, especially those below the mouth that reveal the teeth. The nostrils dilate nervously, producing wrinkles on both sides similar to those caused by laughter. The eyes may be closed tight, to shut out the source of the terror, or they may be wide open, showing the staring whites of the eyes. In that case the eyes are kept wide open by the violent contraction of the orbicularis and the frontalis muscles, which pull the brows and lids upward, forming deep furrows on the forehead.

Even an apparently expressionless face reflects personality. I hope that, using the instructions in this chapter, you will now be able to draw other emotions. I have finished, but you have not. Practice is extremely important, particularly if you plan to take up portraiture or commercial art. Stand in front of the mirror and make sketches of the mouth and eyes in different positions, using yourself as a model.

34

Fig. 34. Anger causes furrows between the eyebrows, wrinkles below the eyes; the mouth opens and the jawbone juts forward.

Fig. 35. For the expression of exasperation the lids open, the eyebrows are raised, and all the muscles around the lips contract.

Fig. 36. Hatred is displayed mainly in the eyes (more open than usual), the furrows of the brows, and the tightly closed jaw.

Fig. 37. Fear means that the eyes are opened wider than usual and the eyebrows are raised, with the mouth tightly opened.

Fig. 38. This expression displays the feeling of terror; the eyes are tightly closed and the mouth wide open.

35

36

37

38

Michelangelo and Laocoön

In Rome, on January 14, 1506, Felice de Freddis was working in his vineyard when an extraordinary thing happened. Felice was a farmer. That day he was digging quite deep when his spade hit something. It was a hard object, but somehow it did not feel like just a stone. "A skeleton?" Felice wondered. He went on digging and found that he was uncovering a statue—he saw arms, chest, head...

News of this find spread rapidly through Rome. "In Felice's vineyard! A Greek statue, eight feet high... several figures... all in marble!" Pope Julius II immediately sent one of his officials, the architect Giuliano da Sangallo. When he arrived at Felice's vineyard, he found that his friend Michelangelo Buonarroti was already there. Michelangelo quietly asked, "Who says that the Greeks did not have a complete knowledge of anatomy?" He approached the statue, pointing out the muscles in the arms, chest, abdomen. "Look, Giuliano, see how the abdominal muscles are contracted, and how the serratus and oblique muscles vibrate."

"It is the *Laocoön*," Sangallo pronounced, "the famous sculpture described by Pliny the Elder, the Roman historian. Wait, Michelangelo, I must dash off and tell the Pope."

Pope Julius II soon arrived. By that time Felice's vineyard was packed with people. At the sight of the Holy Father, they knelt and fell silent. Michelangelo alone did not kneel. Quite oblivious of the Pope's presence, he stood, arms folded, completely absorbed in his study of the sculpture.

While everyone watched expectantly, the Pope came up behind Michelangelo and touched his shoulder gently. Still preoccupied, the painter murmured, "This is man as God created him, in His own image. Glory be to the Creator for ever and ever, world without end." The Pope knelt and replied, "Amen."

The discovery of the *Laocoön* in Felice's vineyard provided new and wonderful inspiration for all the Renaissance artists.

We can imagine Michelangelo going frequently to study the sculpture. At that time he was, among other things, working on the famous statue of Moses. Two years later he started the enormous paintings in the Sistine Chapel.

Michelangelo himself said that his masters were the remains of ancient art found in Florence and Rome. He learned not simply by contemplating them, but by carefully studying their forms, comparing them with his human models, analyzing the body contours and relating this with what he knew of the *inside* workings of the human body.

We shall now try to follow Michelangelo's example and study the position and shape of each muscle in an attempt to understand the bone structure supporting the figure.

Fig. 39. Indeed, it was the sculpture Laocoön, one of the most important works in ancient Greece. Agesandro and his sons Polidor and Athenodor were the sculptors. The theme is based on the myth of Laocoön, the Trojan priest who warned his countrymen to distrust the great horse sculpture that the Greeks had abandoned at the walls of the town —the famous Trojan horse. According to the legend the gods, enemies of Troy took revenge on Laocoön by killing him and his two sons with two huge serpents.

The anatomy lecturer in my art school used to say, "When you can see the skeleton under the flesh, then you can say you know how to draw the human body." Then invariably he would give us a graphic example: "Watch me when I raise my right arm," and he would gravely lift his arm. "Do you know why my right shoulder is higher than my left?" And he would look around at all of us, waiting. If no one answered immediately, he would shout, "Because, when we raise our arm, the clavicle and scapula swing upward. Study the bones. Learn them. *Study!*" And so we did. Eventually we learned the shape, position, and function of all the many bones that make up the human skeleton. You're more fortunate; you can learn them without the terror of class questions and exams.

MUSCLES

1. Temporalis
2. Frontalis
3. Supercilius
4. Orbicularis of the eyes
5. Masseter
6. Levator of the upper lip
7. Zygomatic minor
8. Orbicularis of the lips
9. Zygomatic major
10. Buccinator
11. Risorius
12. Triangularis muscles of the lips
13. Levator menti
14. Sternomastoideus
15. Trapezius
16. Deltoids
17. Pectorals
18. Biceps
19. Pronator
20. Palmaris brevis
21. Palmaris longus
22. Supinator
23. Serratus
24. External oblique
25. Rectus
26. Tensor of the fascia lata
27. Pectineus
28. Abductors
29. Sartorius
30. Rectus of the thigh
31. Vastus internus
32. Vastus externus
33. Patella
34. Gastrocnemius
35. Tibia
36. Tibialis
37. Extensor of the toes

BONES

1. Skull
2. Clavicle
3. Scapula
4. Humerus
5. Spine
6. Ulna
7. Radius
8. Carpal bones
9. Metacarpal bones
10. Phalanges
11. Pelvis
12. Femur
13. Patella
14. Tibia
15. Fibula
16. Tarsal bones
17. Metatarsal bones
18. Toes

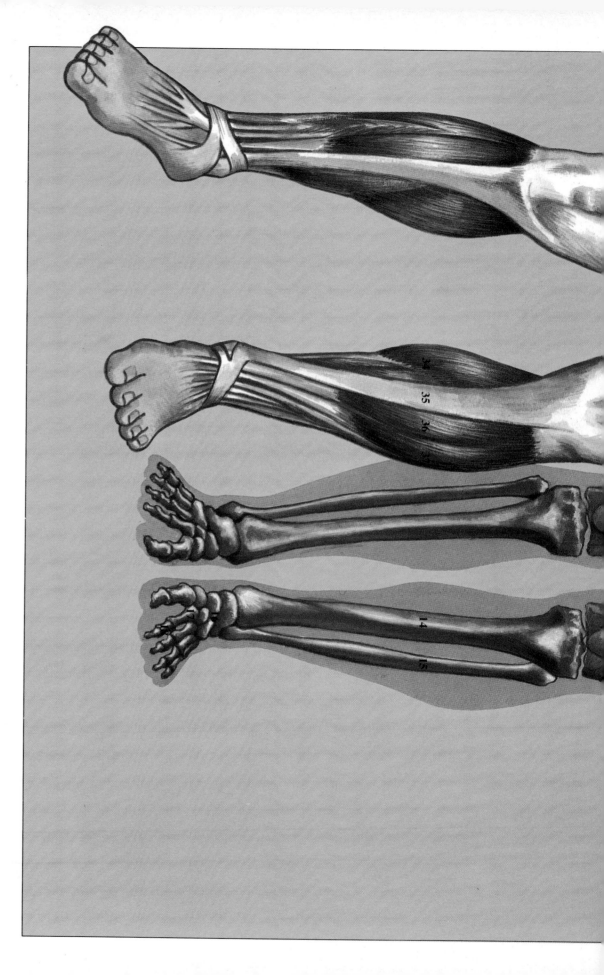

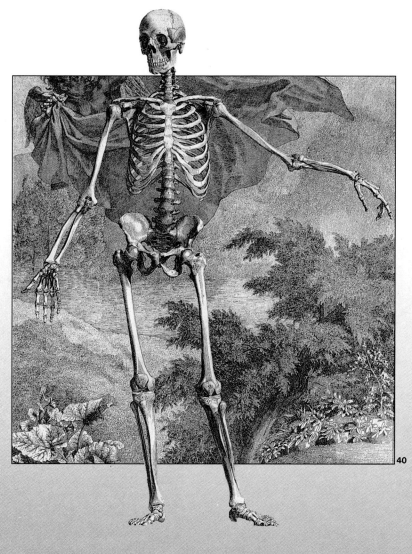

THE HUMAN
—SKELETON—

The skeleton: general principles

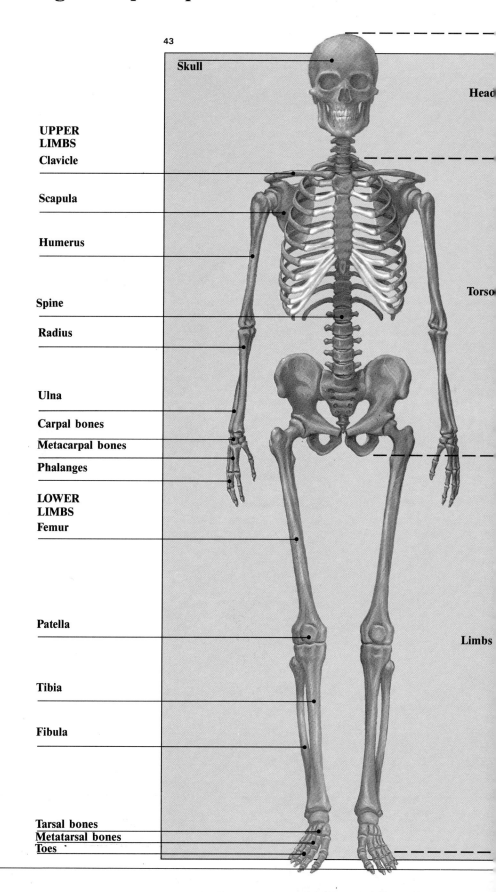

43

Skull

Head

**UPPER
LIMBS**
Clavicle

Scapula

Humerus

Spine

Radius

Torso

Ulna

Carpal bones

Metacarpal bones

Phalanges

**LOWER
LIMBS**
Femur

Patella

Limbs

Tibia

Fibula

Tarsal bones
Metatarsal bones
Toes

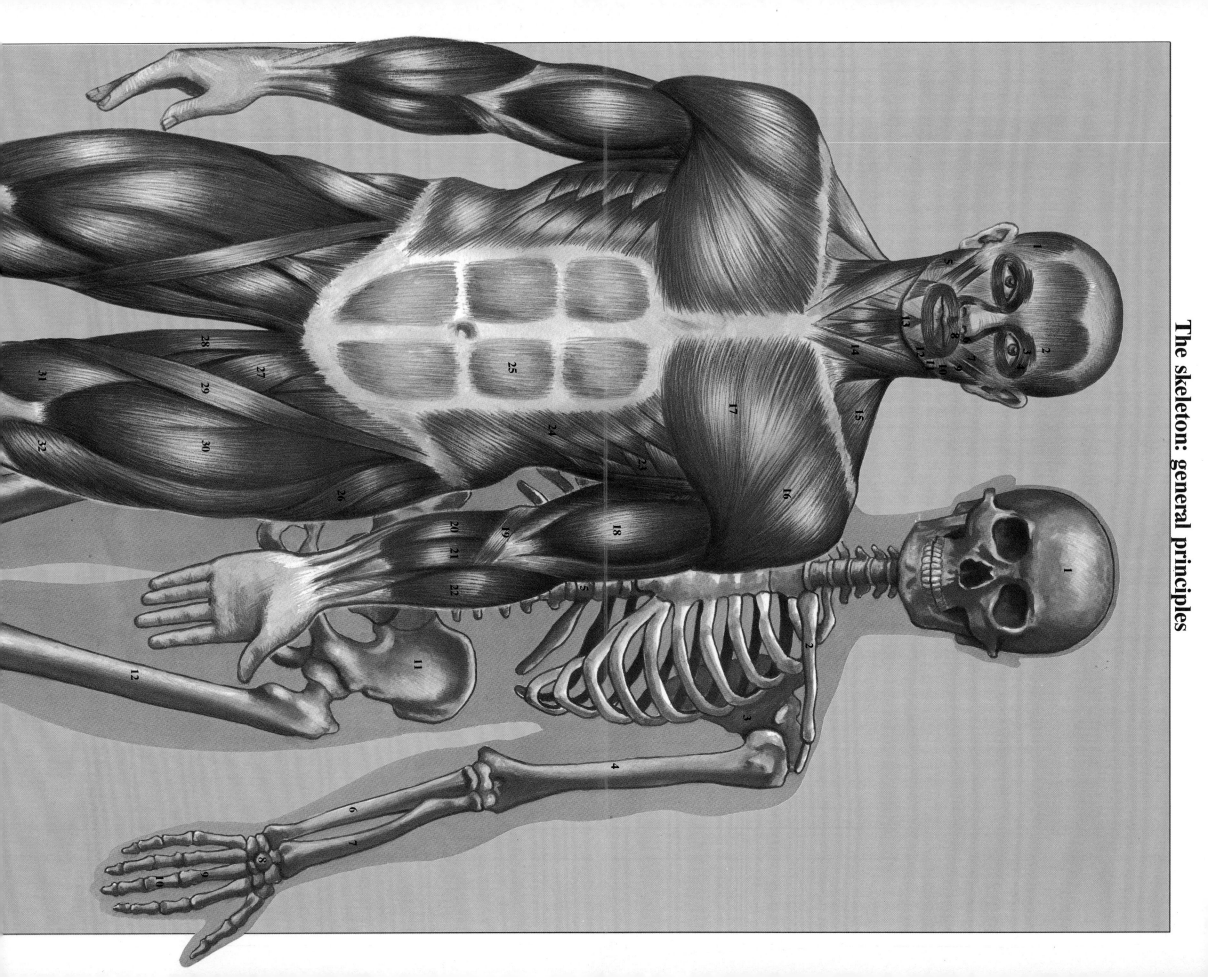

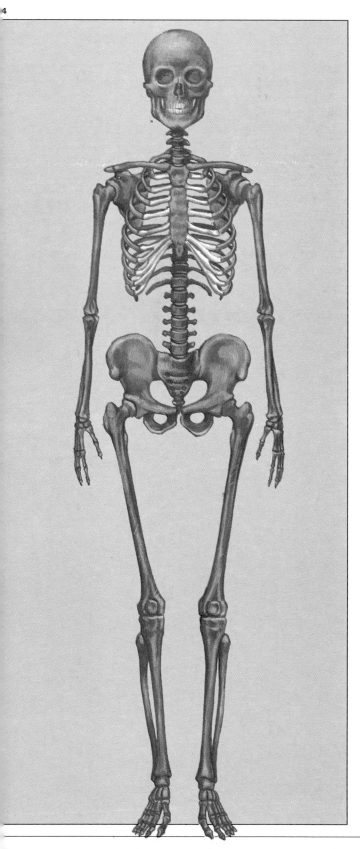

4

The skeleton—the frame of the human figure—is made up of three main sections: head, trunk, and limbs.

The head is formed by the *cranium* and *cervical vertebrae*. The trunk consists of the *spine*, which supports the head, the *thorax*, and the *pelvis*. The upper limbs are linked to the thorax by the *scapula* and we can distinguish the *arm* (with the *humerus*), the *forearm* (comprising the *ulna* and *radius* and linked to the arm by the *elbow*), and the *hand* (made up of the *carpal* and *metacarpal* bones, and the *phalanges*). The lower limbs are connected to the trunk by the *pelvis* and are made up of the *thighs* (with the *femur*), the *calves* (with the *tibia* and *fibula*), and the *foot* (with the *tarsal, metatarsal bones, and toes*). The thigh and the calf meet at the knee joint, where we find the *patella*.

Figs. 43 and 44 show the male and female skeletons. Notice that they are not equal. The male body is taller than the female, has broader shoulders, and narrower hips. A woman has narrower shoulders and wider hips. There are evolutionary reasons for these differences. For thousands of years man has needed physical strength for his work, so his respiratory apparatus and the corresponding bones and muscles, mainly those in chest and shoulders, developed. The woman's wider pelvis evolved because of her child-bearing function, the need for more space during pregnancy. We shall now make a detailed study of each of the parts of the body and the bones mentioned here.

Compared morphology

To help you understand the representation of the human body, compare the dimensions and proportions of the man with the woman, and those of the child with the adult. Notice the numbered segments. Each segment, or module, corresponds to the height of the head for each figure, the tallest boy in Fig. 45 is seven "heads" tall. Using these modules you can keep elements of the body in correct proportion. Study Fig. 46. You see that both the man and the woman have the same basic dimensions: eight units ("heads") high and two units wide. But since the woman's head is smaller, her overall size is smaller, about 4 inches (10 cm) shorter.

See also the male's broader shoulders and the female's broader hips and thinner waist, as well as the lower level of her breasts and nipples.

In childhood and adolescence the growth factors alter the proportional relationship between the different parts of the body.

Fig. 45 shows the three stages of human development, at two, six, and twelve years. Observe how the proportion of various parts of the body shifts. See the greater size of the head of these young males compared with the adult's head. Also note the slimmer shape as the development goes on, and the changes in the proportional relationship between legs and trunk.

45

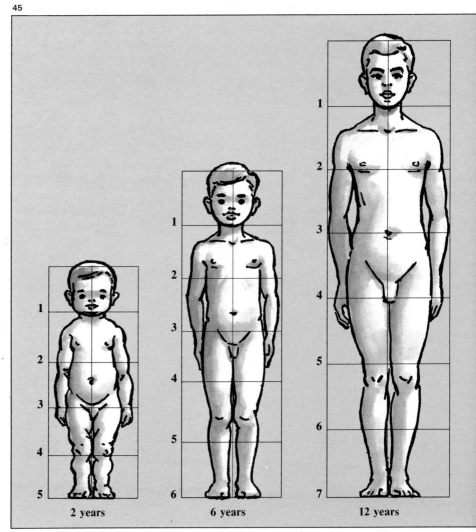

Fig. 45. In a two-year-old child, the head is bigger and the legs are shorter. This is the "square" image of childhood, the pattern for the children painted by artists such as Rubens or Murillo. At six years, the trunk grows faster than the head and the figure is six "heads" high. At twelve the boy is seven heads high and the body structure is quite similar to the adult, except for muscle development.

2 years   6 years   12 years

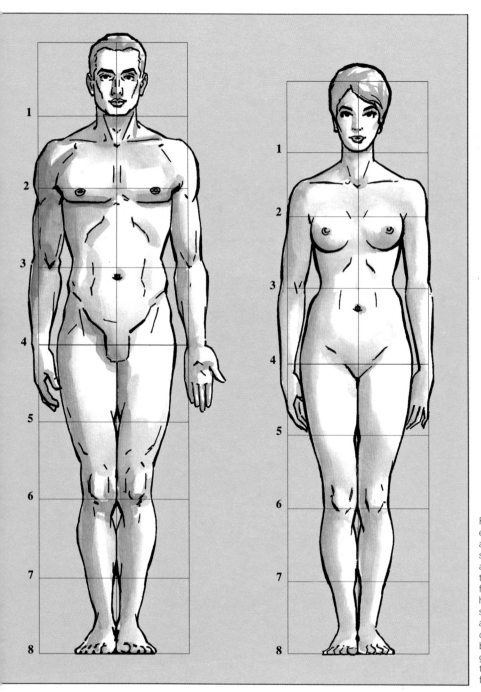

Fig. 46. The main difference between the male and the female is the smaller head in the woman. That is why, though the man is larger, both figures are eight heads high. The man's wider shoulders and the woman's wider hips are key differences, and have to be taken into account to get faithful representations of the human figure.

Articulation of the skeleton

Fig. 47. The mandible is the only mobile bone in the head. The rest are a rigid set with no articulation (except the articulation that links the head to the spine).

The head
You have studied the bones of the head in an earlier chapter (pages 10-13). Remember that it consists of the cranium and facial bones forming a rigid unit except for the mandible, or jawbone, which is mobile and operated by certain muscles when the mouth is opened or closed. (Take a few minutes now to review those pages.)

The trunk
This consists of the spine, the thorax and the pelvis. It is essential that you study this part of the body. When you are fully familiar with the shape and function of the trunk bones, you will be able to draw the body in any posture, because every movement originates in the trunk and it determines the position of the arms and legs. Read these explanations carefully.

The spine
This forms the central axis of the trunk, running the whole length from the base

of the skull behind and in the center of the lower maxillary down to the *sacrum* in the pelvis; the middle section of the spine supports the thorax. It is made up of a column of vertebrae or ring-shaped bones, each of which has a series of extensions, one on each side and one in the center. The central extension, called the *spinous process*, is important for us because in certain positions it is visible just under the skin (Figs. 49A and B on this page).

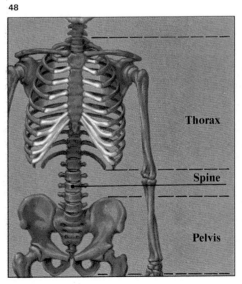

Fig. 48. The trunk is the core of movement in the human body. The essential components are the thorax, the pelvis, and the spine.

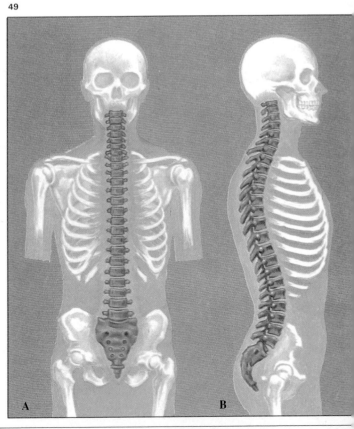

Fig. 49. In a front view the spine appears straight when the body is standing. In a profile view we can see that it is actually curved in an S shape.

When the body is standing straight, the spine lies in the center, forming a straight vertical line when seen from the front or rear; in profile, we can see that it is shaped like an elongated S following the curved shaped of the body (Fig. 49). Within certain limits, the spine can turn twist, and bend forward, backward, sideways. These movements are reflected in the positions adopted by the head, thorax, and pelvis.

In Fig. 50A, for instance, notice that when the weight is placed on the right leg with the left relaxed (known as the *ischiatic position*), the pelvis tilts one way while the thorax tilts in the opposite direction, and the spine twists and straightens out as required. To give another example, Fig. 50B shows a body in the action of walking, with the head and thorax turned to the left and the pelvis to the right, while the spine turns on itself to allow the waist to move and the neck to turn.

An important reminder: the movements of the spine are visible on the surface and in the back a deep hollow appears that corresponds exactly to the shape of the spine (Fig. 51).

Notice in Fig. 50C that the seventh vertebra becomes prominent; every human body shows this when the head is bent. Test it yourself. Bend your head, feel your spine. It's easy to locate, isn't it?

Fig. 50. Here we have three examples of the different movements of the spine: the ischiatic position (A), the walking attitude (B), and the standing position (C) with the head bent to show the spinous process of the seventh vertebra.

51

Fig. 51. It is easy to note the movement of the spine on the surface of the back. The turning movement reveals the characteristic hollow where the spine is.

50

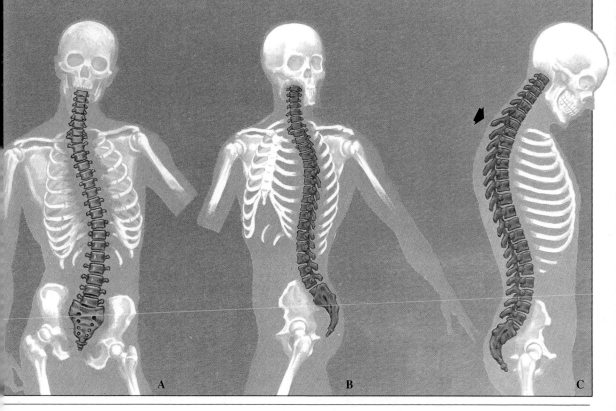

A   B   C

The trunk skeleton: the thorax

Ischiatic position

This position takes its name from the balancing movement of the *ischium* bone (in the middle lower part of the pelvis), which bends to one side or the other depending on the hip movement. As the body weight is shifted to one leg, the pelvis and the thorax shift in counterpoint. A person walking or standing in a relaxed way displays this position (see Figs. 50 and 52B).

The thorax

This consists basically of a series of arches, the ribs, joined to the spine at the shoulder and in the front to the *sternum bone* (Fig. 53). The sternum marks an axis line in the chest and is of vital importance in the structure of the human figure. Fig. 54A shows us the thorax in profile; notice how it influences the shape of the chest, emphasizing point A, the end of the sternum, which is usually visible on the surface. Also notice the position of the spine in relation to the thorax. The thorax includes a set of joints at the ends of the ribs that enables it to move freely as the torso twists and turns.

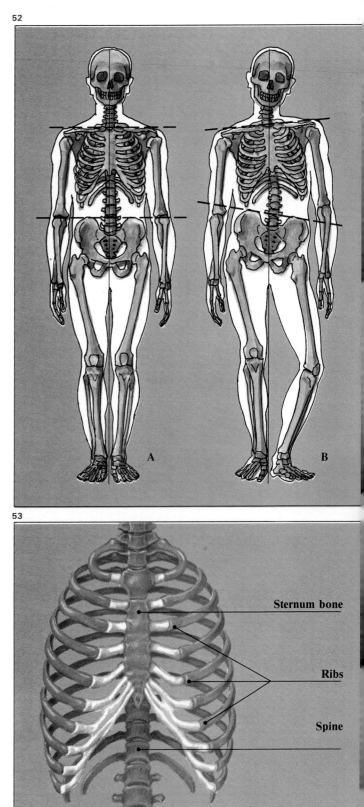

52

A   B

53

Sternum bone

Ribs

Spine

Figs. 52 A and B. Study these illustrations of the skeleton. In the standing position, with the weight equally distributed on both legs, the axis of the hips is perpendicular to the spine (A). If the weight is supported on just one leg and the other leg is relaxed (B), the pelvis lifts on the side that supports the weight and the thorax tilts in the opposite direction.

Fig. 53. The movement of the thorax is determined by the movement of the spine and the joints at the ends of the ribs. These joints allow the thorax to stretch when the lungs fill with air.

54

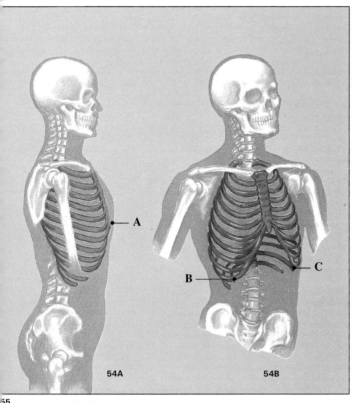

54A   54B

For example, when a deep gulp of air is taken in, the lungs stretch and the arch of the ribs lifts a little along a horizontal line, which makes the outline of the ribs visible on the surface, and also the lower arch of the thorax (the pit of the stomach) in which the peaks marked B and C become prominent (Figs. 54B and 56).

Note that these points are also visible when the torso is bent backward with the arms raised and the abdomen tensed, tightening the waist and stretching the chest. Finally, look at Fig. 55, the torso of Laocöön, which shows a superb example of the outline of the thorax with the axis of the sternum fully visible: it also demonstrates the outline of the ribs and the pit of the stomach.

Fig. 54. When the thorax stretches, the outline of the ribs is visible on the surface (A), as well as the lower arch of the thorax forming the pit of the stomach (B and C).

Figs. 55 and 56. Compare the torso of Laocöön with a real figure. Although the shape of the statue is ideal, we can identify the contour of the sternum, the lower arch of the thorax, and the ends of the ribs.

55

56

The trunk skeleton: the pelvis

Figs. 57 and 58. The pelvis is formed primarily by the sacrum, which joins to the spine and the hip bones, which join to the femurs. The upper end of the hip bone is the iliac crest.

Figs. 59 to 61. We can easily appreciate the iliac crest in these three positions of the human body: the ischiatic position (Fig. 59), with the leg place to the rear (Fig. 60), and with the torso leaning backward (Fig. 61).

Fig. 62. These are the bones that work in the articulation of the arms. The clavicle is placed on the upper end of the shoulder and connects to the sternum and the scapula.

The pelvis

This is formed primarily of three bones: the *sacrum*, which ends in the *coccyx* (a kind of rudimentary tailbone), and the two hip bones (Figs. 57 and 58). The top of the pelvis is joined to the spine and the bottom to the thigh bones (*femurs*) (Fig. 60). As we have already mentioned, the male pelvis is narrower than the female (Figs. 43 and 44).

To the artist, one of the most important points of the pelvis is the *iliac crest* forming the hip (see Figs. 57 and 58), which is clearly visible on the human body. You can feel this bone on your own body by pressing your thumb on your hip. In thin people this bone is very prominent and determines their contours at waist level when seen from the front. It can also be seen in almost all bodies when they adopt certain postures like the ones illustrated on page 49. Fig. 59 shows the outline of the iliac crest (marked with an arrow) when the pelvis is swung to one side and

the weight rests on one leg (the position shown in Figs. 50A and 52B). The same occurs in Fig. 60, where the body is shown from behind with the right leg placed to the rear, and also in Fig. 61, which gives a three-quarter view with the torso leaning backward. Examine these figures to see the shape of the pelvis when viewed from the front, from the rear, and in a three-quarter frontal view.

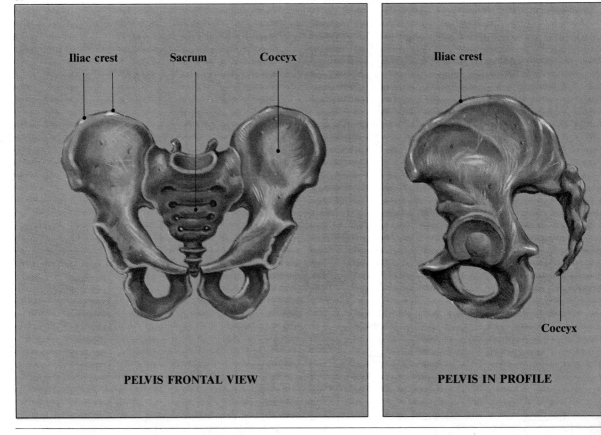

57

Iliac crest   Sacrum   Coccyx

PELVIS FRONTAL VIEW

58

Iliac crest

Coccyx

PELVIS IN PROFILE

The clavicle

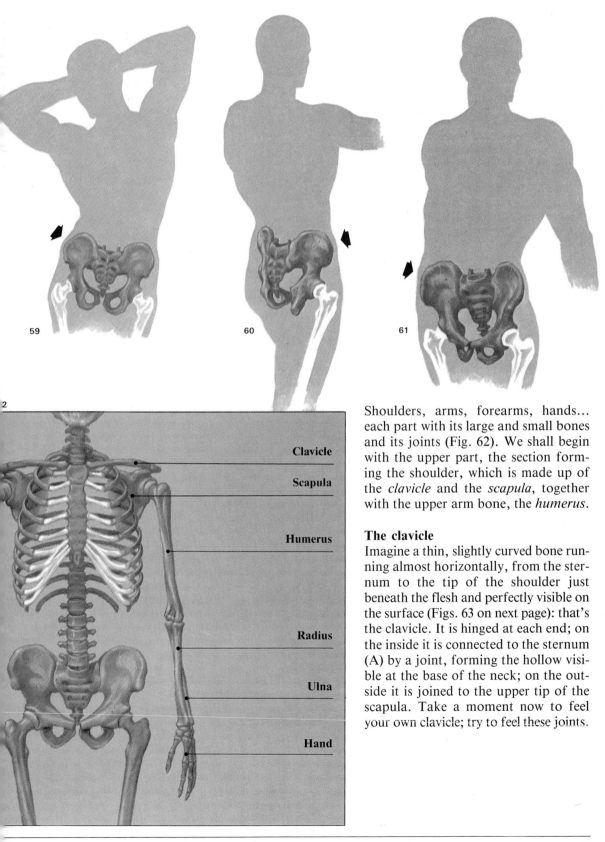

59

60

61

2

Clavicle

Scapula

Humerus

Radius

Ulna

Hand

Shoulders, arms, forearms, hands... each part with its large and small bones and its joints (Fig. 62). We shall begin with the upper part, the section forming the shoulder, which is made up of the *clavicle* and the *scapula*, together with the upper arm bone, the *humerus*.

The clavicle
Imagine a thin, slightly curved bone running almost horizontally, from the sternum to the tip of the shoulder just beneath the flesh and perfectly visible on the surface (Figs. 63 on next page): that's the clavicle. It is hinged at each end; on the inside it is connected to the sternum (A) by a joint, forming the hollow visible at the base of the neck; on the outside it is joined to the upper tip of the scapula. Take a moment now to feel your own clavicle; try to feel these joints.

The upper limbs: bones of the arm

Fig. 63. The clavicle joins the sternum forming a hollow visible at the base of the neck (point A). You can notice the hollow if you feel your clavicle with the fingers.

Figs. 64 and 65. The scapulae are visible on the surface of the back because of their characteristic wedge shape on the rear side of the shoulders.

Figs. 66 and 67. The articulation of the shoulder is very complex: The sternum is linked to the clavicle, the clavicle joins the scapula, and the scapula is linked by a joint to the humerus.

The scapula

As you can see (Fig. 64) the scapula is a triangular bone shaped to form a smooth convex surface... rather like a pair of patches stuck to both sides of the back. It is important for the artists to remember the following points: (a) the angle of the inside lines shown in Fig. 64; (b) the wedge-shaped ends, which can nearly always be seen on the surface, especially in thin bodies (Fig. 65); (c) the position and distance between the bones (notice that they are symmetrical and they change position when the arms are moved, as we shall see later); and (d) the crest along the upper section of each bone, rising slightly toward the sides of the body. (Check these points in Fig. 64 and also study the surface appearance and shape in Fig. 65).

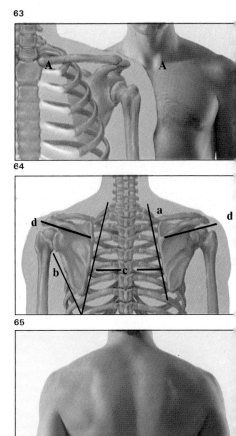

63

64

65

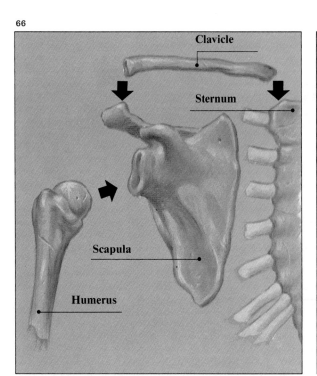

66

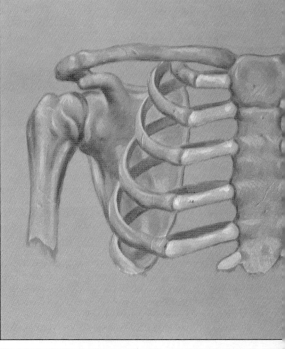

67

...e humerus

...is is the only bone in the upper arm; ...starts at the shoulder and ends at the ...ow (see its shape and position in Fig. ...).

...e clavicle, scapula, and humerus are ...connected by joints and move in con- ...nction when the shoulder is raised or ...wered or the arm raised or stretched: ...ch of these actions produces different ...ositions of the bones in relation to each ...her. We have to study this carefully if ...e want to make an accurate drawing. ...e shall start by studying in detail the ...ticulation system of all these bones in ...gs. 66 and 67.

The clavicle is linked by joints to the sternum.

The scapula is linked by a joint to the clavicle.

The humerus is linked by a joint to the scapula.

...hen the arms are raised, the clavicle ...ings upward. Its inside end is con- ...lled by its joint with the sternum (Fig. ...). The scapula swings outward from ...e point where it joins the clavicle ...ig. 69).

When the arms are stretched forward, the clavicles form an angle whose peak lies at the junction with the sternum (Fig. 70); simultaneously, the scapulae move apart toward the arms (Fig. 71).

When the arms are stretched backward, the bones move in the opposite direction; the clavicle also swing slightly backward with their axis still on the sternum (Fig. 72), and the scapulae move close together in the middle of the back (Fig. 73).

Figs. 68 to 73. Note the positions of the clavicle, the scapula, and the sternum when different movements take place. The contours and hollows are noticeable on the surface.

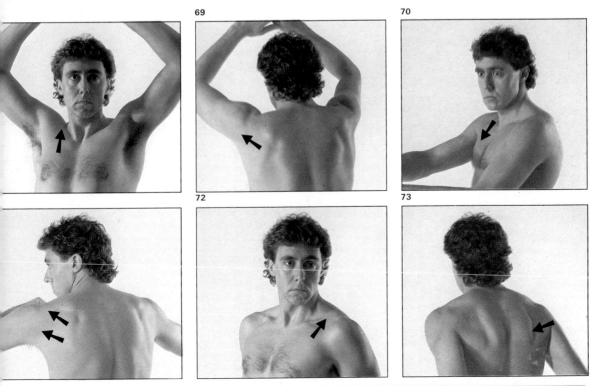

69

70

72

73

The upper limbs: skeleton of the forearm

Figs. 74 and 75. The shape of the elbow responds to the system of joints at that point: the end of the humerus (visible inside and outside the elbow), and the olecranon of the ulna, which forms the point of the elbow.

Figs. 76 and 77. The movements of the ulna and the radius enable us to twist the forearm: these movements are called pronation (when the hand is turned with the palm upward) and supination (palm downward).

Ulna and radius

It is extremely important to study the forearm bones (*ulna* and *radius*) and their movements at elbow and wrist. First of all, observe the shape of the elbow when the arm is bent (Fig. 76). Notice the three bony points that become very apparent on the surface: one, on the inside of the elbow, is the tip of the *humerus* (A); the other, in the middle at the base, is the *olecranon of the ulna* (B) (notice that this forms the point of the elbow); and the third, on the outside, is the other side of the tip of the *humerus* (C). Study the shape and position of these projections in Figs. 74 and 75. Now, look at the system of joints that enable us to twist our hands, a movement in which the ulna and radius have a vital function. These bones are parallel in Fig. 76 (the *supination* position) while in Fig. 77, where the hand is turned with the palm downward (the *pronation* position), the radius is forced to cross over the ulna. Finally, look at the hand in Fig. 77 and notice how the pronation position makes the end of the ulna quite visible on the surface.

76

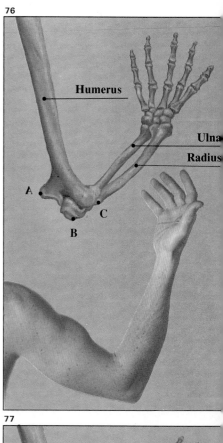

Humerus

Ulna

Radius

A

C

B

77

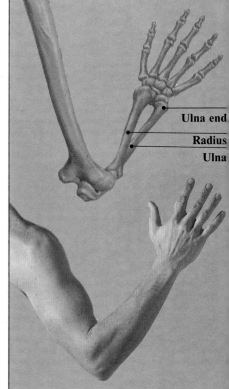

Ulna end

Radius

Ulna

74

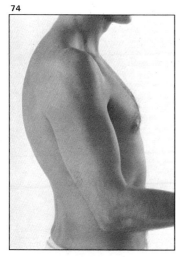

75

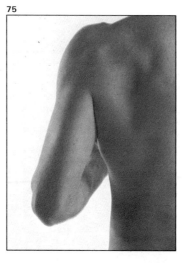

The bones of the hand

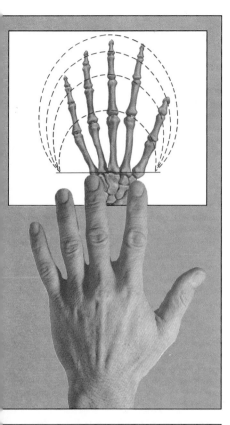

The hands

Our hands are made up of the *carpal* and *metacarpal* bones and the *fingers*, which in turn are divided into three groups known as the *phalanx, second phalanx,* and *third phalanx*. The carpal bones form the wrist. The five metacarpals provide the back and palm of the hand; they are articulated in the wrist and the fingers. Three articulated bones form the fingers, but the thumb has only two (Fig. 78). Most of the bones in the hand are visible on the surface. It is essential to study the exact shape of these bones, especially the fingers. Note, for instance, that they are thinner in the middle than at the ends: remember this when you are drawing a bony or thin hand. Notice, too, that the finger bones are simply an extension of the metacarpals; this means that the joints of the fist must always be in a straight line with the knuckles (Figs. 79 and 80). Finally, an important detail, which we shall encounter again when we actually start to draw hands: *the joints in the fingers correspond to a series of curved lines* (Fig. 78).

Figs. 78 to 80. The articulations in the hands are neatly visible on the surface. It's a good idea to study them carefully. The apparent complexity of this system is nothing but the normal relationship between metacarpal bones of the palm and phalanges (the bones of the fingers).

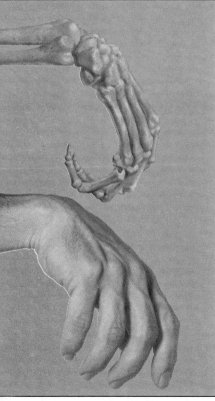

80

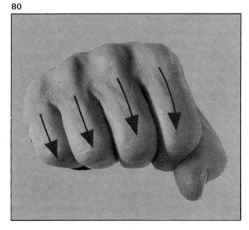

The lower limbs: bones of the thigh

The lower limbs start in the thighs with the *femur*, from which extend the *patella* in the knee, the *tibia* and *fibula* in the leg, and finally the foot, which consists of the *tarsal* and *metatarsal* bones and the *toes* (Fig. 81).

The femur

This is the longest and bulkiest bone in the body. It is thick, strong, and very mobile (although less so than the humerus), because of the way it is joined to the pelvis. Look at this system carefully (Fig. 82); the end of the femur is shaped like a knob and fits into a spherical cavity in the pelvis. From this knoblike attachment there is a short extension toward the outside of the limb—called *the neck of the femur*—which then forms another bony protuberance in which the *trochanter major* is most prominent; this is the part that takes the shock when we fall on our side. The trochanter major influences the outline of this part of the body. In Fig. 83 notice the prominence of the hip bone (A); the smooth hip cavity just below that (B), caused by the absence of bones near the surface; and another smooth protuberance (C) produced by the *trochanter major*.

The main part of the femur is long, thick, and almost cylindrical, but not entirely straight. In profile you can see that it is very markedly curved, which produces the curve on the front of the thigh (Fig. 84).

Finally, when it reaches the point where it connects with the patella, the femur again becomes wider, causing the bony protuberances of the patella on both sides of the knee: this is what we will study next.

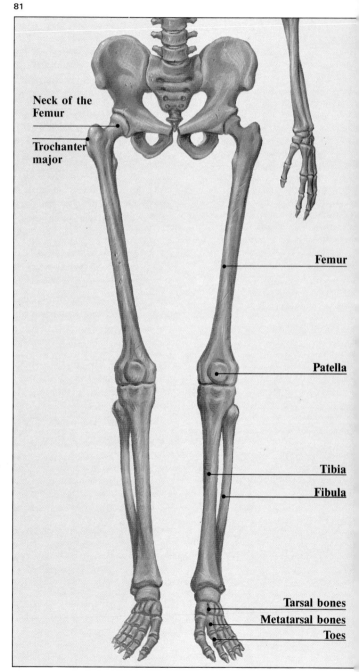

81

Neck of the Femur

Trochanter major

Femur

Patella

Tibia

Fibula

Tarsal bones

Metatarsal bones

Toes

Fig. 81. The lower limbs are joined to the pelvis by the neck of the femur. The femur is joined to the tibia and the fibula by the knee. And the tibia and the fibula are linked to the foot at their lower end.

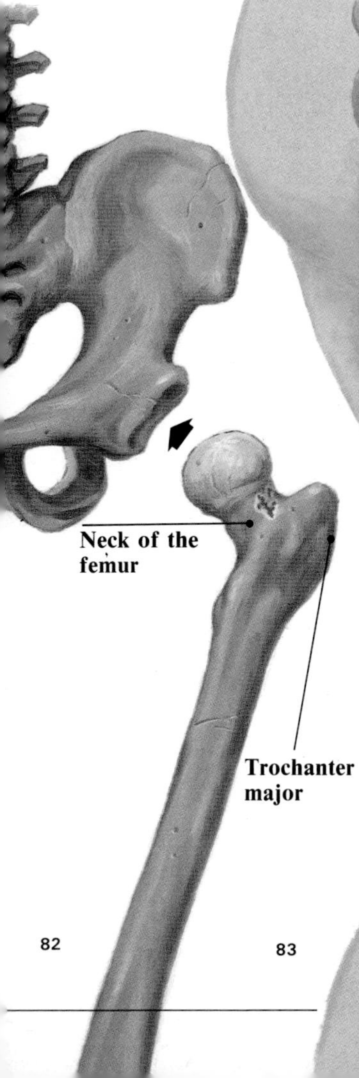

Fig. 82. The femur is linked to the pelvis by a knoblike attachment at the end of the neck of the femur, which fits into a spherical cavity in the lower part of the pelvis. The upper protuberance of the femur is the trochanter major.

Fig. 83. The articulation of the femur with the pelvis is noticeable at the lower outside end of the hip, when the leg is stretched backward. Some important points are visible: the iliac crest (A), the trochanter major (C), and clear hollow on the hip (B) between both protuberances.

Fig. 84. The femur is the longest and one of the strongest bones of the skeleton. Its curved shape determines the external appearance of the thigh in a profile view.

A

B

C

C

Neck of the femur

Trochanter major

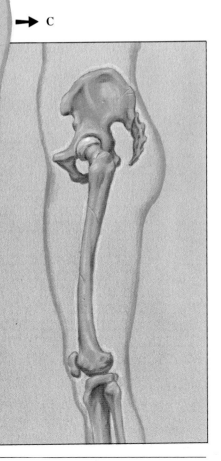

82

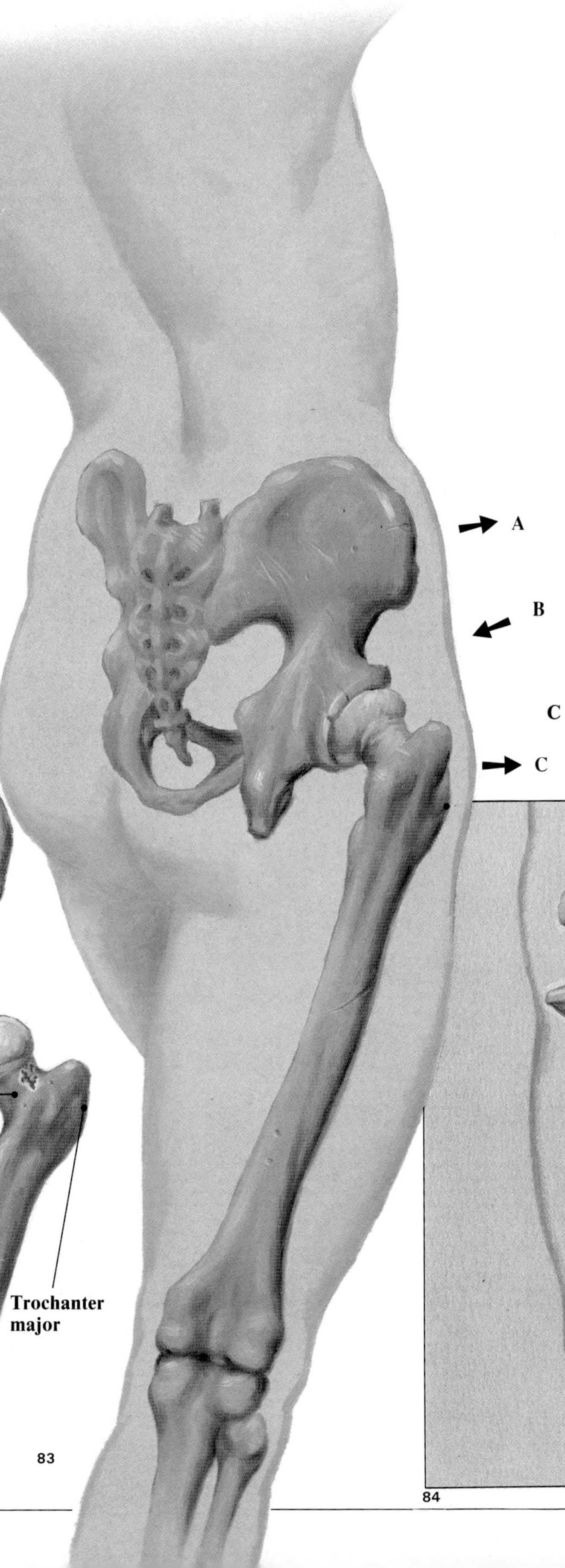

83

84

The lower limbs: bones of the leg and the foot

Patella, tibia, and fibula
The feet

Study the positions of the lower leg bones in Fig. 85. Notice that the *patella* (the kneecap) is shaped like a small shell and, although not attached to any other bones, it is controlled by special ligaments and muscles. When the leg is straight, the patella lies above the joint in front of and overlapping the femur so that it is visible on the surface (Fig. 86). When the leg is bent, the patella enters the cavity produced by the movement of the joints. Note that the end of the femur and the tip of the tibia are now in different positions in relation to each other (Fig. 87). Study carefully the shape of the knee joint in both positions.

The whole length of the tibia can be seen on the surface, as Fig. 88 shows. Notice the position and shape of the bones forming the knee when seen from the front and observe how they affect the visible shape of the knee. We can also see the protuberance caused by the tip of the *fibula*. At the lower end, the leg bones form projections known as the *malleoli* (the ankles). It is worth remembering that the inside malleolus lies somewhat higher than the outside one (Figs. 88 and 89).

Finally, the shape and position of the *tar-*

Figs. 85 to 87. In the articulation of the knee the femur, the tibia, the fibula, and the patella all work together. The articulation of the femur with the tibia and the fibula may emphasize the shape of the patella when the leg is stretched, and diminish it when the leg is bent.

85

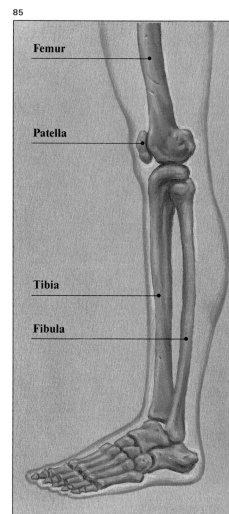

Femur

Patella

Tibia

Fibula

86

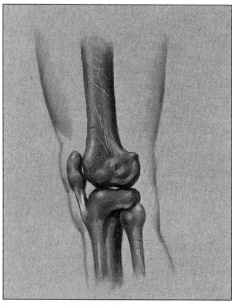

87

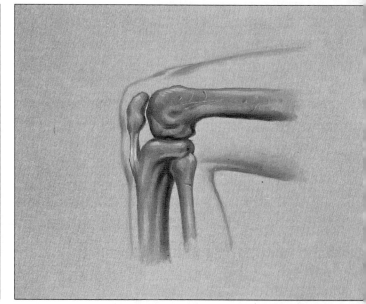

sal, metatarsal, and *toe* bones in the foot correspond to those found in the hand, but we should point out the arched shape of the foot and the protuberance forming the heel, which is caused by the *os calcis.* Note that the shape of the foot is largely conditioned by its bone structure (Fig. 90).

Fig. 88. Compare these two representations of the legs. It is not difficult to identify the bony protuberances: the patella, the tip of the tibia, the tip of the fibula, and the malleoli.

Fig. 89. Observe the articulation of the ankle. The inside malleolus lies somewhat higher than the outside.

Fig. 90. The articulations of the foot are as complex as the articulations of the hand, but they are not so easily visible on the surface. Nevertheless, the protuberance of the os calcis and the toe bones are obvious.

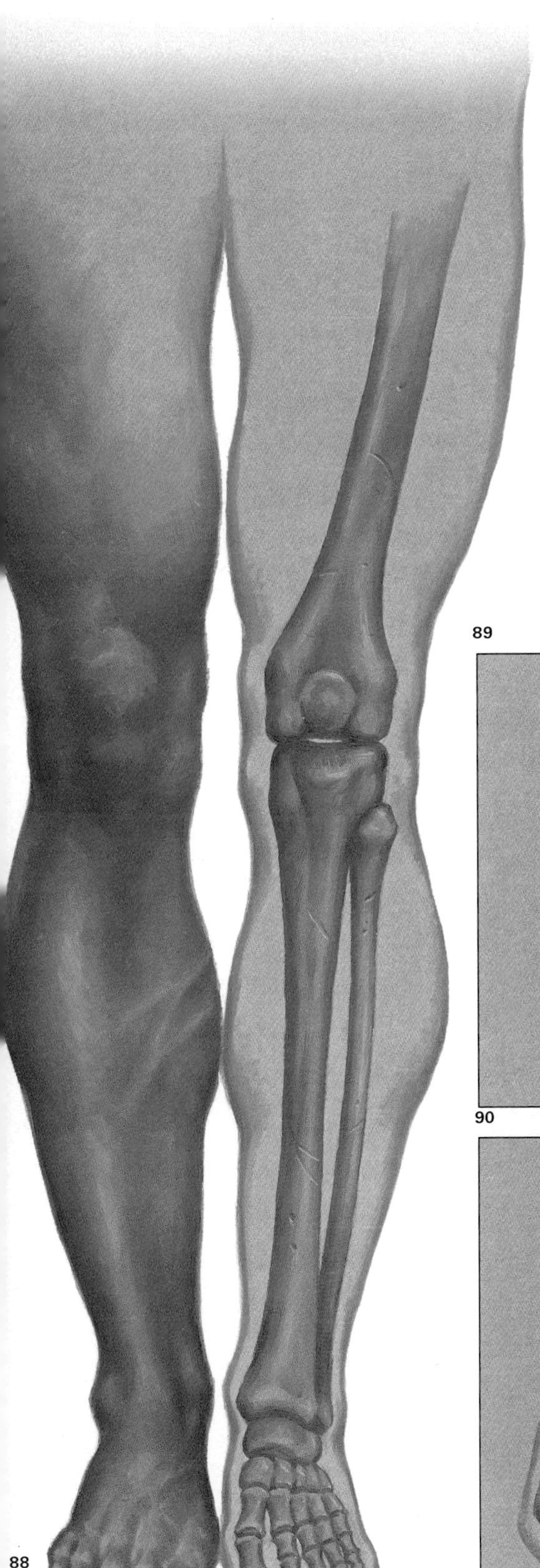

89

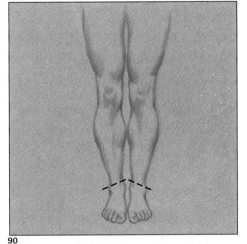

90

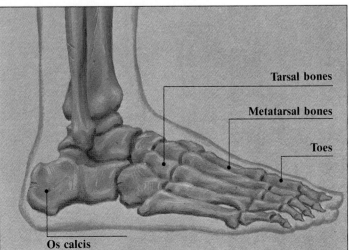

Tarsal bones

Metatarsal bones

Toes

Os calcis

88

"When a dog wags its tail, or a child tries to stand up, or when you scratch your nose, a series of processes are produced which are so complicated that, in comparison, the mechanism of the H-bomb is simplicity itself." These are the words of J.D. Ratcliff, M.D., in an article entitled "The Mysterious Operation of the Muscles."

The muscle mechanism that controls our movements is amazing indeed, but fortunately, we do not need to go into all the complex scientific details here. We will study the position and shape of the muscles and analyze their functions solely from the artist's point of view. You will, for example, learn that the external appearance of an arm is shaped by its muscles and that its position—raised, stretched, folded—is created by the movements of the muscles with the bones and their joints. In brief, you will learn how to draw the human body because you have studied its form and movements.

To start with, look at the general principles on page 60; *they apply to the entire muscular system.*

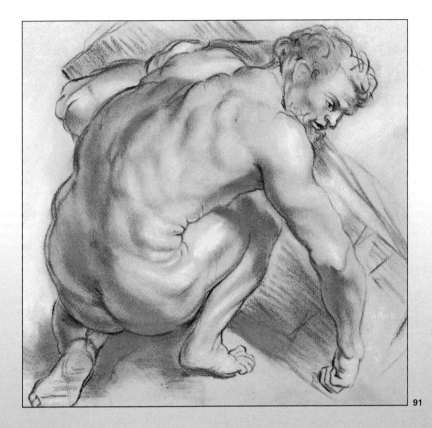

91

MUSCLES
—OF THE—
HUMAN BODY

General principles of the muscular structure

1. Muscles are the organs that activate our body movements through their ability to contract and relax, as if they were made of rubber.
2. Muscles are divided into two groups:
   a. *Flat muscles*, which contract slowly and involuntarily, like those that operate the intestine during the digestive process.
   b. *Striated muscles*, which contract quickly and voluntarily; we move them deliberately when we walk, raise our arm, open our mouth, clench our fist. These are the ones we artists have to study.
3. Generally speaking, muscles are made of bunches of fibers that terminate in one or more tendons linking the muscles to the bones.
4. The shape of the muscles varies in keeping with the function or movement they operate, and they can be classified as:
   a. *Circular*, ring-shaped for closing bodily openings.
   b. *Orbicular*, loop-shaped, like those that close the eyes and mouth.
   c. *Flat and broad*, like the forehead.
   d. *Fan-shaped*, like the temporal muscles that operate the jawbone.
   e. *Spindle-shaped*, bulky in the middle and slender at the end, like those that move the arms.
5. *Spindle-shaped* muscles are the most common type in the human body. A familiar is the *bicep* on the arm; clench your fist and feel it contract.

We know that muscles are the "motors" of the bones; what we need to know now is how they work to move the bones. Most of the body's bone movements are controlled by the *lever principle* of physics, where two different forces—power and resistance—work against a fixed point, the fulcrum. Figs. 92 and 93 show the arm movement as an example of this lever principle. Note the bones and muscles involved in this movement: in the arm the humerus bone is controlled by the bicep and tricep muscles, which are in front of and behind it, respectively. The fulcrum is situated in the elbow, where the humerus meets the two bones of the forearm: the radius and the ulna. Notice that the lower tendons of the two muscles are joined to the bones at that fulcrum point.

Fig. 91. Rubens, great master of the human form, often chose poses that stressed muscle activity. In this picture by the Flemish master, art lies in the deep and certain knowledge of human anatomy.

Figs. 92 and 93. The muscular activity involved in bending the arm is based on the lever principle. The point of support (fulcrum) is the elbow; both bicep and tricep muscles can exert two powers: the first, to raise the forearm; the second, to relax it.

92

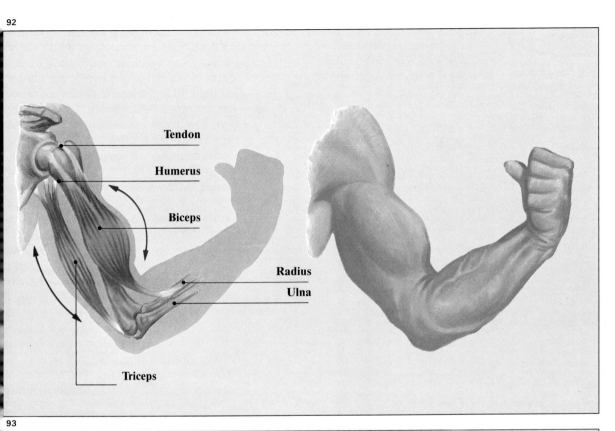

Tendon

Humerus

Biceps

Radius

Ulna

Triceps

93

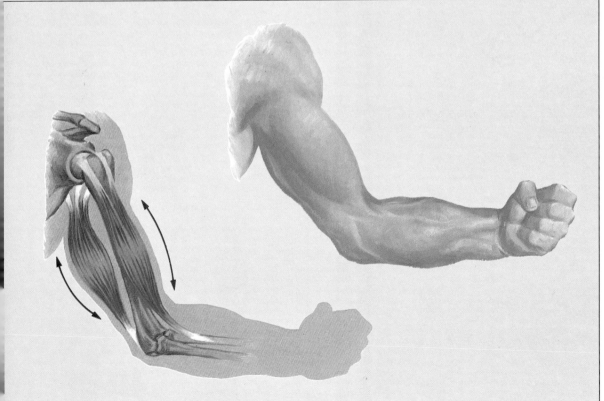

Anatomy of the muscles

Figs. 94 to 97. A thorough knowledge of human anatomy is necessary to lend the drawing of the human figure the exact contours of the body, even if the proportions are correct (Figs. 94 and 96). A good drawing needs not only the right proportions and dimensions, but also a perfect representation of the shape and surface look of the human body (Figs. 95 and 97).

"I think I know how to draw a properly proportioned figure," a young artist said, "with each part of the body in its right place, but sometimes I don't know exactly how to give form to an arm or leg by means of shading."

This is a very common problem. A beginner's work frequently resembles the badly drawn examples in Figs. 94 and 96. Many have only a rough idea of the shape of the human body; they know, of course, that there is a round muscle in the arm and a protuberance on the knee and have read somewhere that the human body is "a combination of cylinders" and here we have the result. Compare these to Figs. 95 and 97, the same figures properly drawn, using a knowledge of anatomy.

You can manage drawings like this; it isn't difficult, but it does require both study and practice. Remember:

The surface look of the human body is mainly determined by the special shape of each muscle.

Bear in mind—and, if necessary, re-read—our study of the muscles of the head and face in the earlier chapter. Now we are ready to study the muscles of the rest of the body.

94

95

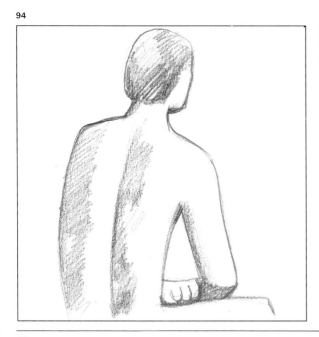

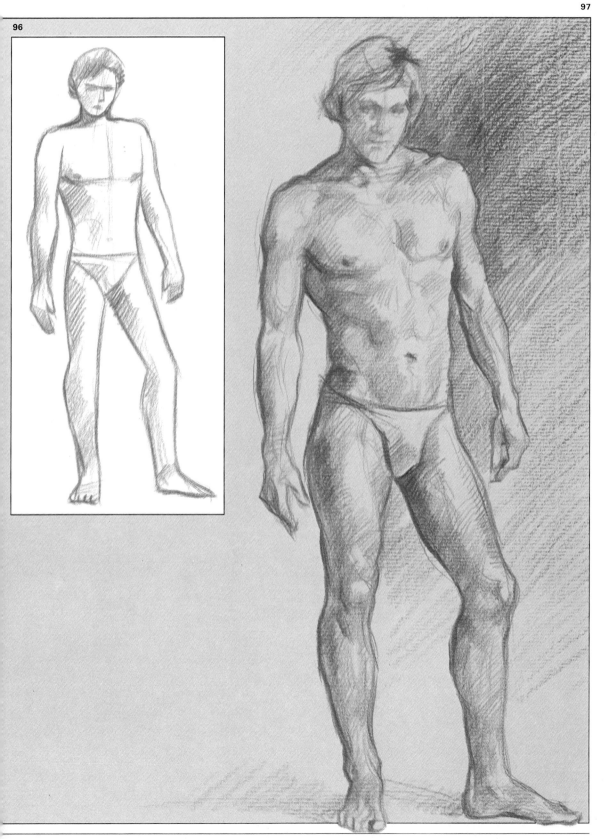

Muscles of the human figure (front view)

Sternomastoideus

This neck muscle starts at the base of the skull behind the ear, with one section running down to the sternum and the other to the inside end of the clavicle. Two matching muscles create a V shape in the neck. When it contracts, this muscle lowers the head and may simultaneously bend it or turn it to one side if its counterpart on the other side contracts. In the latter position the muscle is almost vertical. In either case it is clearly visible on the surface and may emphasize the small triangular cavity formed by its two lower sections (A in Fig. 98). We should also mention the prominence of the Adam's apple or larynx in the middle of the throat.

(The neck contains fourteen other muscles, but we shall not deal with these here since they do not show on the surface. Fig. 99 also shows the *trapezius*, which we shall study later.)

Deltoids

These form a kind of cap over the end of the shoulder. We use them to move the shoulder forward and backward and especially to raise it; they also help to stretch and lift the arm backward. Every arm movement affects the shape of the *deltoids* in close harmony with the shape adopted by the *pectorals*. The depression marked E is clearly visible on muscular bodies (Fig. 98).

Pectorals

This bunch of muscles covers the chest. It forms a strong, short tendon that is controlled by the humerus bone underneath the deltoid muscle; this makes it possible to lower our raised arms.

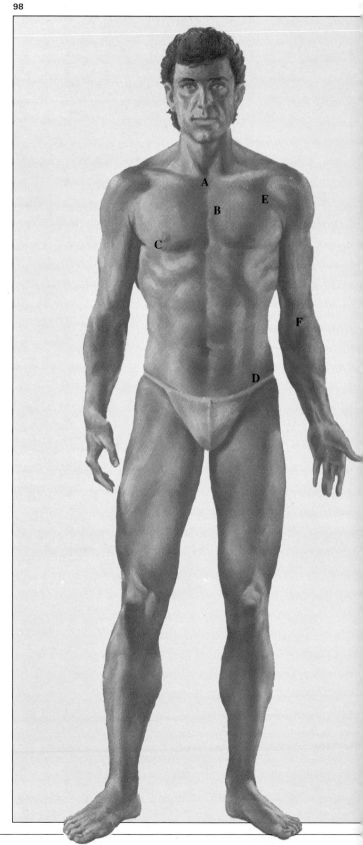

98

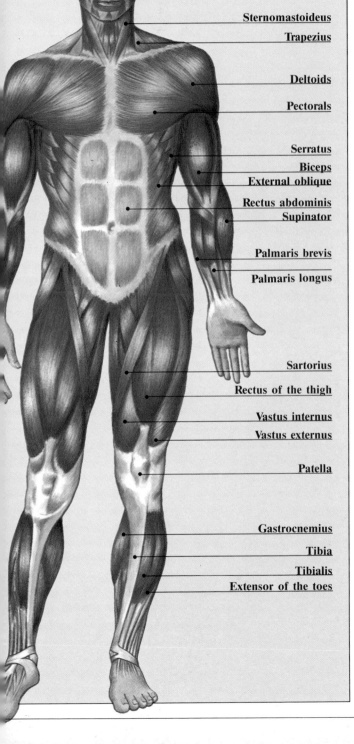

Sternomastoideus

Trapezius

Deltoids

Pectorals

Serratus

Biceps

External oblique

Rectus abdominis

Supinator

Palmaris brevis

Palmaris longus

Sartorius

Rectus of the thigh

Vastus internus

Vastus externus

Patella

Gastrocnemius

Tibia

Tibialis

Extensor of the toes

It also helps to extend the arm forward. The projection of this muscle in the middle of the chest helps to form the furrow running through its center (B in Fig. 98) and also fixes the horizontal line that completes the shape of both sides of the chest (C). When the arms are raised it almost disappears as the ribs become more prominent; when the arms are extended with the hands clasped together, the pectoral muscle is more noticeable.

Serratus
This name is derived from the serrated or sawlike form of this muscle. It lies below the pectoral and next to the armpit and is used to bring the scapula forward and lift the shoulder blade; it also helps to lift the ribs. It can be seen on muscular people on both sides of the thorax; the top section is very evident, less so lower down.

External oblique
This is found on both sides of the thorax, stretching down to the hips, and enables the thorax to adopt special positions in relation to the pelvis. In the lower section near the hips and groin, it forms a smooth horizontal protuberance (D).

Rectus abdominis
This consists of three or four pairs of rectangles along the center of the abdomen and stomach: the last of these divisions coincides with the navel. It is used to bend the thorax over the pelvis and vice versa and can be seen when the figure is standing upright.

Muscles of the arm and forearm (front view)

Biceps

The bulky, smooth shape of these muscles can be seen on the front surface of the arm even when relaxed. They are used to bend the arm and also help to place it in the supination position (palm up; see page 52) or to raise it. This is a very powerful muscle.

Pronator

This lies obliquely across the inside of the forearm near the elbow joint. As its name suggests, it turns the arm to the pronation position (palm down). Its upper section is particularly obvious and, in conjunction with the supinator and the end of the biceps, it forms the V-shaped fold or depression on the surface of the joint (F in Fig. 98, page 64).

Supinator

This places the forearm in the intermediary position between pronation and supination and is primarily used to bend the arm, an action that clearly brings out its shape.

Palmaris brevis and palmaris longus

These two muscles give the front and middle of the forearm their shape and become apparent when the hand is bent back over the forearm. In certain positions, as when clenching the fist, the tendons at the ends of these muscles stand out on the inside of the wrist.

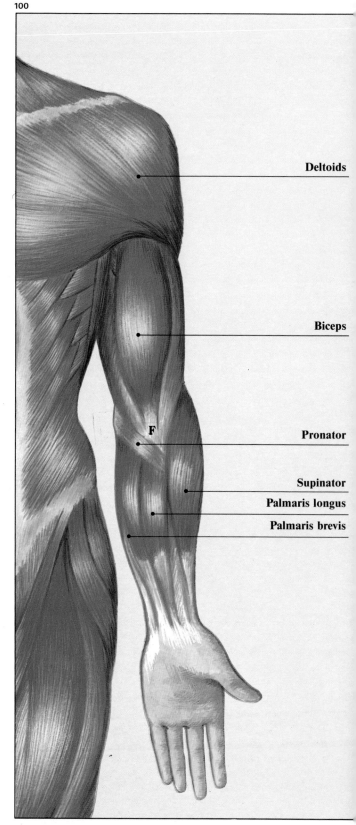

100

Deltoids

Biceps

F

Pronator

Supinator

Palmaris longus

Palmaris brevis

Fig. 100. The deltoid is the larger muscle of the arm and it controls the articulation of the shoulder. On the front surface of the arm is the bicep, a very powerful and obvious muscle. The pronator turns the hand and forms the V-shaped fold on the inner surface of the elbow joint. If we clench our first while we raise our forearm, we can distinguish the shape of the supinator, palmaris minor, and palmaris major.

Radial

This lies in the forearm beside the supinator and controls the hand when it bends backward. It is so near the supinator that usually we see only one muscle, but it can be distinguished in muscular arms and when the hand lifts a heavy object, when the radial bends the forearm back toward the shoulder.

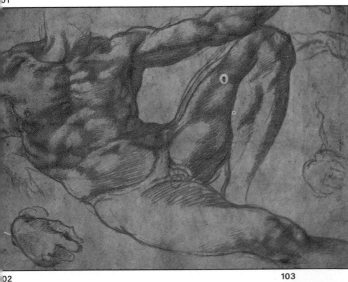

Fig. 101. Michelangelo Buonarroti, *Study on Adam*, British Museum, London. Michelangelo was a genius not only as a sculptor but also as a painter. He often sketched torsos; note the detail here as well as the strength of this drawing technique; all the muscles can be identified.

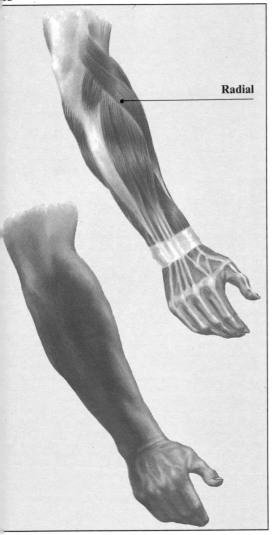

Radial

Fig. 103. Peter Paul Rubens, *Study for a Crucifixion*, British Museum, London. Muscle tightness is noticeable in this figure, especially in the arms, which support the body weight.

Muscles of the thigh and leg (front view)

Pectineus and adductors
These cover the inside upper section of the thigh near the groin. They do not have much effect on the shape of this part of the body.

Rectus of the thigh
This starts at the hip and ends in a powerful tendon connected to the tibia; it is one of the muscles that activate the thigh.

Vastus internus
This appears on the inside of the thigh near the knee.

Vastus externus
This lies on the outside of the thigh near the knee. In combination the *rectus, vastus internus* and *vastus externus* form the shape of the thigh when seen from the front. The vasti act together mainly to raise the leg (to kick, for instance). When contracting to perform such movements, the three muscles merge and are visible only as one mass. When they relax, however (which means that the leg, too, is relaxed and bent), the lower ends of the three muscles appear on the surface above the knee (G in Fig. 98 on page 64). If the leg is stretched to produce a powerful contraction of the muscles, these projections become more evident but are located slightly higher, like the patella (H in Fig. 98).

Sartorius
This is a long strip of muscle running obliquely across the front of the thigh from the hip to the inside of the knee (level with the top of the tibia). It flexes the thigh over the pelvis when one leg is placed over the other. Its shape produces a smooth furrow on the thigh, extending obliquely through the combination formed by the *rectus* and *vastus internus* (Fig. 99, page 65).

Gastrocnemius
We shall study this when we examine the rear view, but please observe its shape and position in Fig. 105.

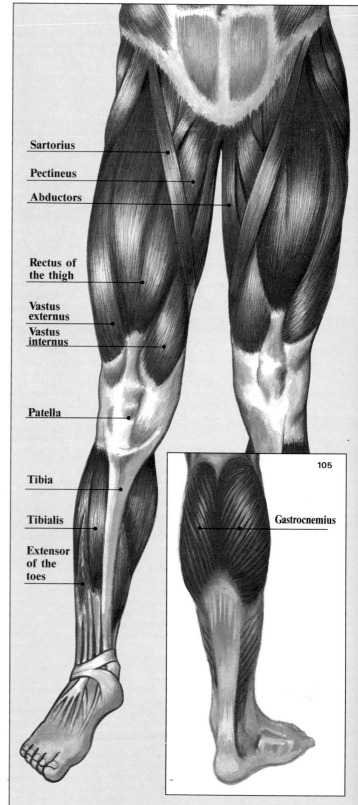

104

Sartorius

Pectineus

Abductors

Rectus of
the thigh

Vastus
externus

Vastus
internus

Patella

Tibia

Tibialis

Extensor
of the
toes

105

Gastrocnemius

Fig. 106. Théodore Géricault, *Nude Warrior with a Spear*, National Gallery of Art, Washington. Although the work of Romantic painter Théodore Géricault is far away from that of the Renaissance, he shows a strong interest in anatomy. The model's pose is relaxed but the muscular structure can be easily appreciated. Observe the characteristic shape of the vastus externus and the gastrocnemius when the leg is stretched, the volume of the adductors as the other leg bends, the contour of the deltoids and the biceps on the right arm, and the protuberance of the supinator on the left arm.

Tibia
This is a bone, not a muscle. We show it (and the patella) in Fig. 104 so that you will remember how close to the surface it is.

Tibialis
This muscle lies beside the tibia on the outside of the leg and runs from the patella to the foot, which it can bend upward. On some bodies you can see the furrow separating this muscle from the *extensor of the toes.*

Extensor of the toes
This is found on the outer side of the leg and, as its name suggests, it moves and extends the toes. At the lower end it forms a number of tendons, which are usually visible and branch off to the toes.

Muscles of the human body (rear view)

Trapezius
The whole of this muscle forms an irregular quadrilateral (known in geometry as a trapezoid, from which it gets its name) on the upper part of the back between the neck and shoulders. It is also visible from the front (Fig. 99). The trapezius can contract entirely or only partially. When partially contracted it can raise the shoulders, bend the head to one side, raise the scapula, and so on. When the whole muscle operates, it moves the scapula toward the middle of the back, where its lateral and upper edges become prominent.

Infraspinatus and teres major
These two muscles cover the triangular area formed above and on both sides of the scapulae. They operate the humerus and also work in extending the arm inward and backward. The infraspinatus is particularly obvious since it produces a marked protuberance when the arm is moved backward or rotated. The contractions of the trapezius, infraspinatus and teres major very rarely hide the surface outline of the scapulae (A in Fig. 107).

Dorsal muscles
Notice the shape and position of the dorsal muscles. Also observe the light-colored diamond-shaped area in the middle of the lower back (A in Fig. 108). This is called the *aponeurosis*. It is a flat, muscular tendon instead of the more usual ropelike tendon. (You can see the latter type in Fig. 108 at the bottom of the rectus of the thigh, where it joins the patella.)
The dorsal muscle activates the humerus, to allow the arm to be lowered or stretched backward. It is thin at the aponeurosis and thick near the ribs, producing at the armpit a long dorsal protuberance that is extremely evident (B in Fig. 107) particularly when the arm is raised. When the body is bent backward the dorsal contracts, bringing out the *aponeurotic line* (line B in Fig. 108).

107

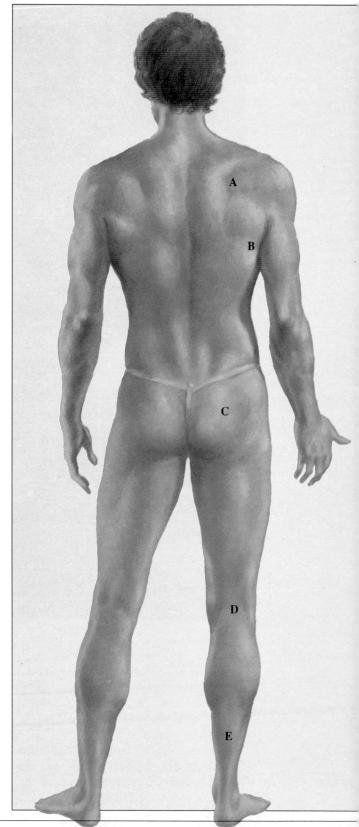

8

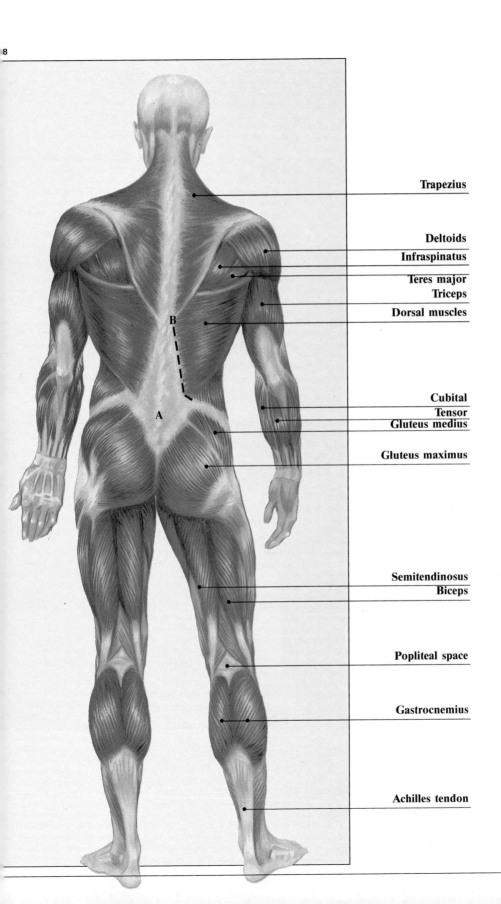

Trapezius

Deltoids
Infraspinatus

Teres major
Triceps
Dorsal muscles

B

Cubital
Tensor
Gluteus medius

A

Gluteus maximus

Semitendinosus
Biceps

Popliteal space

Gastrocnemius

Achilles tendon

Muscles of the lower limbs (rear view)

Gluteus medius

This is found in the upper part of the buttocks, determining the shape of both the buttocks and the hip. A smooth protuberance occurs at this point.

Gluteus maximus

This is the bulkiest muscle in the buttocks and, like the medius, it becomes evident during certain movements of the femur, when it stretches the leg or turns the femur outward. It is relaxed or partially relaxed when the body is standing still (C in Fig. 107, page 70). When the body is bent backward and the muscles contract with some violence, the buttocks change shape by stretching in a lateral direction. The gluteus muscles provide the balance for the body when it is upright.

Semitendinosus and biceps

These are the muscles that shape the rear of the thigh. They play the same part as the biceps in the arm: enabling the lower leg to bend back under the thigh. When the leg is bent they contract and become very evident on the surface. This contraction emphasizes the two end tendons inserted at both sides of the patella. These tendons separate and then become linked with the gastrocnemius near the fold in the knee joint, producing a rhomboid-shaped cavity known as the *popliteal space*, which is clearly visible from the rear (D in Fig. 107).

Gastrocnemius

The calf muscles begin as separate units in the popliteal space and then combine to form a long, powerful tendon called the *Achilles tendon* —the thickest and most powerful tendon of the entire body— which in turn is connected with the os calcis or heel. The gastrocnemii are the muscles used for movement, since they control the foot and raise the body when we walk, run, or jump. Their bulky, rounded shape is characteristic of the calf and emphasizes the depression formed at the bottom end of the muscle, where it joins the *Achilles tendon* (E in Fig. 107).

109

Fig. 109. Théodore Géricault, *A School*, National Gallery, London. ''School'' means the painting or drawing practice that aims at the right realization of the techni- cal and anatomic fundamentals. Try to identify every muscle on the back, the arm, and the leg of the figure. Most of them are clearly visible.

110

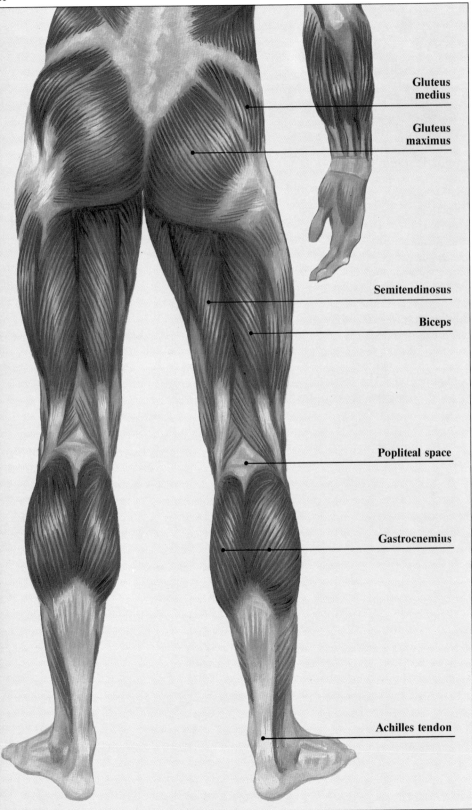

Gluteus medius

Gluteus maximus

Semitendinosus

Biceps

Popliteal space

Gastrocnemius

Achilles tendon

Fig. 110. The legs muscles on the rear side start in the lower back. We can distinguish the gluteus medius and maximus. The semitendinosus and biceps have a function similar to the biceps in the arm. Two tendons link these muscles with the gastrocnemius, producing a cavity known as the popliteal space. The bulky, rounded shape of the gastrocnemii is characteristic of the calf. These muscles join the Achilles tendon (the longest in the body), which is connected with the os calcis or heel.

Muscles of the body in profile

One muscle in particular needs careful examination when seen in profile.

Tensor of the fascia lata
This lies on the outside of the thigh, beginning in the iliac crest and ending in the tip of the tibia (A in Fig. 111). It tightens the *aponeurosis femoral* (also called the *fascial lata*) and helps to flex and raise the thigh. It also helps maintain balance when the body is standing on one leg. It is very conspicuous, especially when the knee is bent; in fact, it very rarely relaxes completely.

This completes our examination of the principal muscles of the human body. Study the descriptions carefully, memorizing their shape, position, and function.

I know there are a lot of them, but don't forget that I have omitted quite a few so as not to overwork you. Study the texts and illustrations until you feel thoroughly familiar with them, and then go on to the following practical exercises.

111

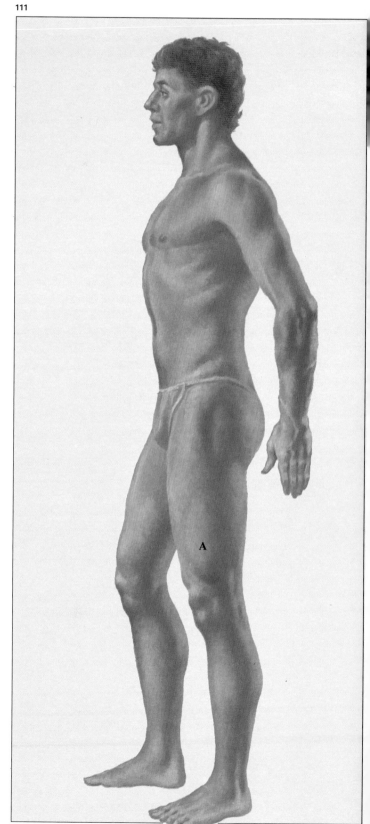

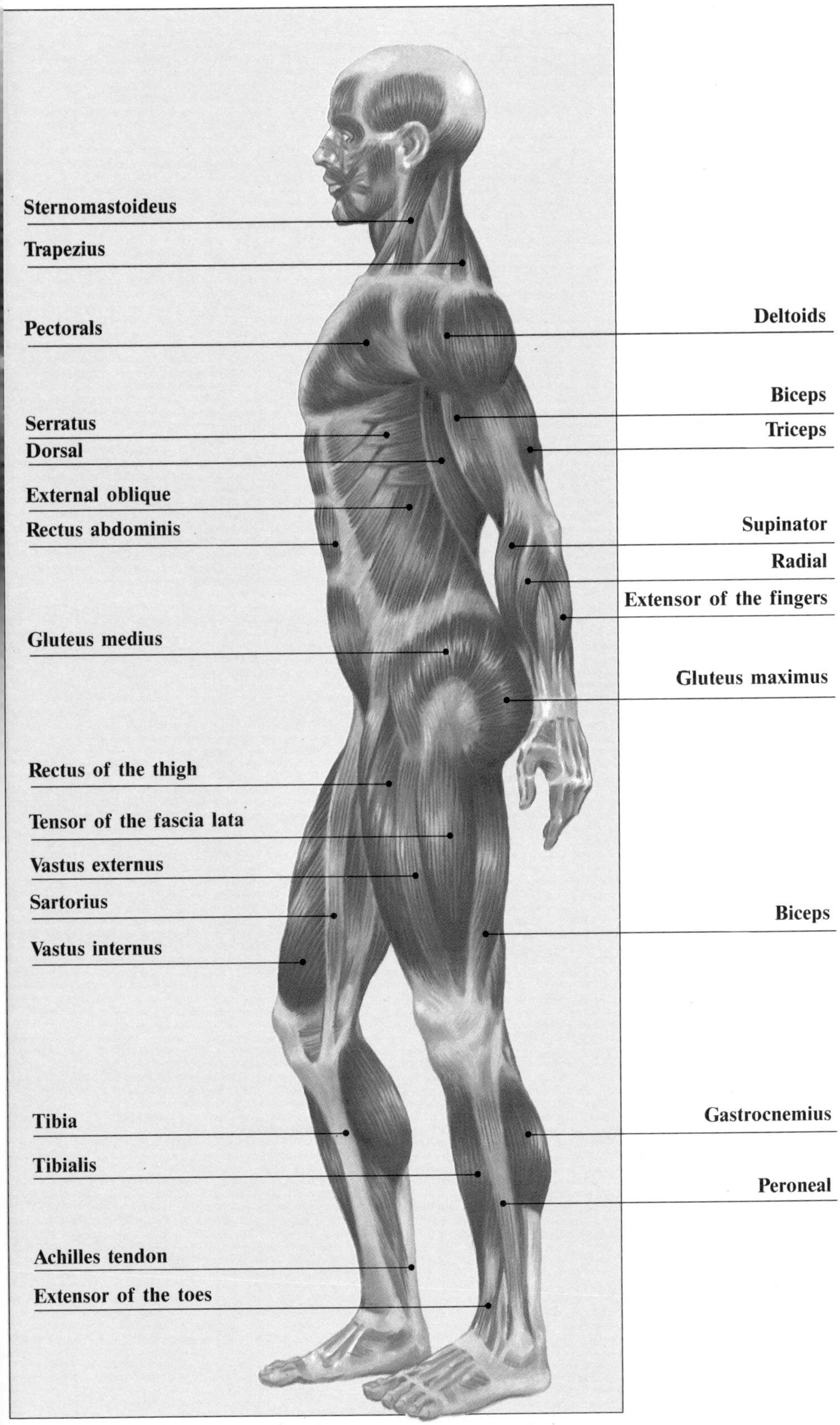

Sternomastoideus

Trapezius

Pectorals

Serratus
Dorsal

External oblique

Rectus abdominis

Gluteus medius

Rectus of the thigh

Tensor of the fascia lata

Vastus externus

Sartorius

Vastus internus

Tibia

Tibialis

Achilles tendon

Extensor of the toes

Deltoids

Biceps
Triceps

Supinator

Radial

Extensor of the fingers

Gluteus maximus

Biceps

Gastrocnemius

Peroneal

Two examples by the masters

113

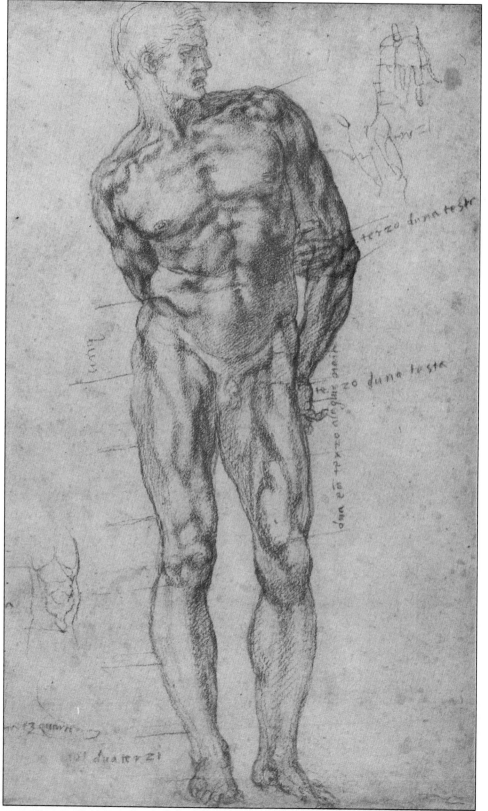

Fig. 113. Leonardo da Vinci, *Male Nude*, Royal Library, Windsor Castle, London. Leonardo was a pioneer in the study of human anatomy. His drawings have been a pattern of realistic representation for centuries, not only for artists but also for medical students. This figure shows the level of scientific faithfulness he aimed at. This is one of the best models to use as practice in figure drawing.

- 114

Fig. 114. Giulio Romano, *Two Figures*, Royal Library, Windsor Castle, London. Leonardo's faithfulness of anatomic representation inspired other artists. Romano is one of them: these figures present a clear and elegant muscular structure and articulation of the movements. Try to reproduce these drawings and you will admire them even more.

Now you know the
complex system of
muscles that is under
our skin, its several
shapes and functions.
To make this
knowledge useful
in art, we must apply
it to drawing. In this
chapter you will find
different exercises based
on the many muscle
groups you have
studied. As we work
through these exercises
together, you will come
to appreciate more fully
how important a
knowledge of anatomy
is to strong, lifelike
drawing.

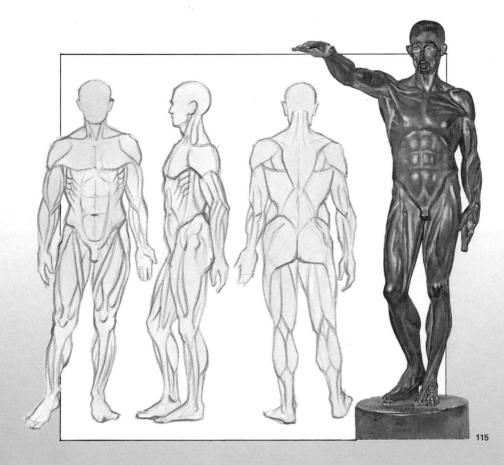

115

HUMAN
—ANATOMY—
IN PRACTICE

"The Flayed": frontal and profile views

116

Here we have a frontal view (Fig. 116) and a profile view (Fig. 118) of the sculpture "The Flayed", by Houdon, a common sculpture in the art schools since the late eighteenth century. Its main feature is that it clearly shows all the muscles of the human anatomy. Do two drawings, concentrating on the muscles, like the sketches in Fig. 117. First sketch in the general outline of the figure to get the overall proportions correct, and then add the lines created by the individual muscles. Study the shape and position of every muscle. See, for instance, in the front view, the shape the rectus of the thigh and the vastus internus and the vastus externus adopt: In the right leg, which is slightly contracted, they almost merge in one single long, rounded protuberance; in the left leg, they are relaxed and the three muscles show clearly, over both vasti. Notice also the special shape that the right pectoral adopts when the arm is raised.

Look at the position and shape of the sternomastoideus in the neck and, in the outstretched arm, the V-shaped hollow formed by the end of the biceps and the start of the supinator and palmares. Notice the depression formed where these two muscles unite in the forearm. Can you see the smooth ribbonlike shape of the pronator? Examine the protuberances of the serratus, and the depression formed by the arch of the ribs. Look at the shape of the fascia lata tendon shown in the thigh and the conspicuous evidence of the biceps in the calf, curving inward to join the Achilles tendon.

Figs. 116 to 118. This is the famous sculpture of a flayed male figure by Jean Antoine Houdon. The sketches for these figures are reproduced on figure 117. Try to draw them, considering both the statue and the sketch, so that you can clearly appreciate the distribution of muscles. Your drawings should resemble these. Avoid modeling with shadows to get volume, since it would interfere with clarity and detail. Just try to place the muscles in the right spot.

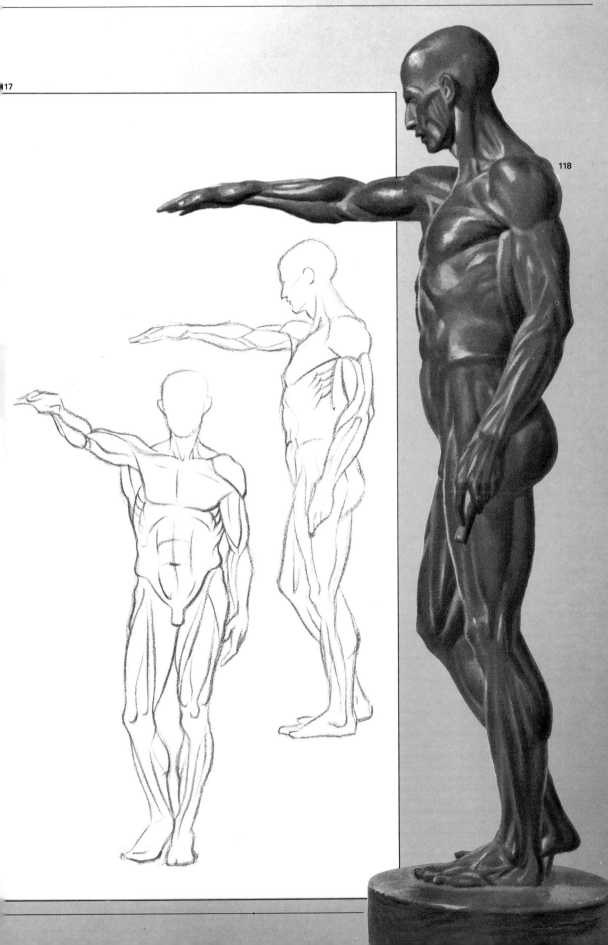

"The Flayed": three-quarter and back views

Figs. 119 and 120 show three-quarter and back views. Notice how the scapula stands out when the arm is raised and stretched forward (remember that the clavicle does more or less the same); the shape of the scapulae and the hollow or axis of the spine become evident. Can you see the edges of the trapezius meeting in the hollow of the back? Below them you see the outline of the dorsal muscle; look at the lower limit of the deltoids on the left shoulder. Look down at the leg and notice the tendons of the biceps and the semitendinous forming the popliteal space, which is such an important feature of the back of the leg. Are you beginning to feel familiar with this wonderful combination of muscles, tendons, and aponeuroses? Of course you are! I told you it wasn't difficult.

119

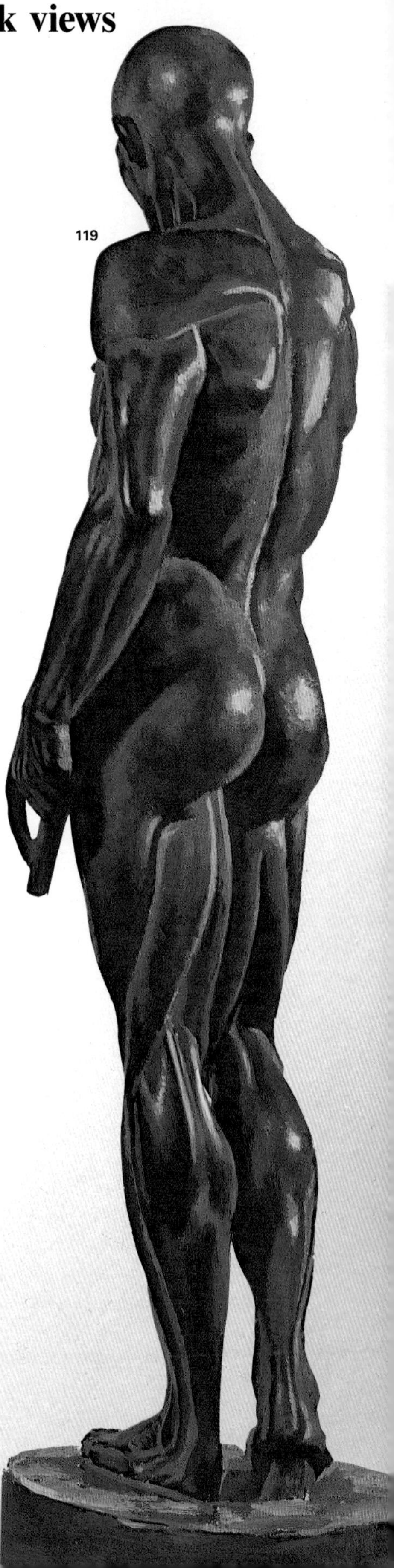

Fig. 119. In this three-quarter back view, you can appreciate the protuberances of the gluteus muscles. When the tensor of the fascia lata flexes the thigh, the gluteus muscles relax and their volume is less noticeable. Notice, too, the os calcis linked to the Achilles tendon.

Fig. 120. Study the special look of the shoulder on the back when the arm is raised. The deltoids, trapezius, infraspinatus, and teres major create hollows and ridges along the scapula. Notice the hollow that forms on the upper part of the spine between the two trapezius muscles.

120

121

Fig. 121. As in the previous exercise, try to place the different muscles in the right positions.

Muscles in activity: two instances

Figs. 122 and 124. These illustrations show different points of view of the same position. Here all the muscles are tight. In figure 122 we can distinguish the serratus, the external oblique, and the rectus abdominis on the torso; the deltoids, the triceps, the supinator, and radial on the shoulder and arm; and gluteus muscles, rectus, tensor of the fascia lata, sartorius, and vastus internus on the legs. Figure 124 presents all the muscles in a front view.

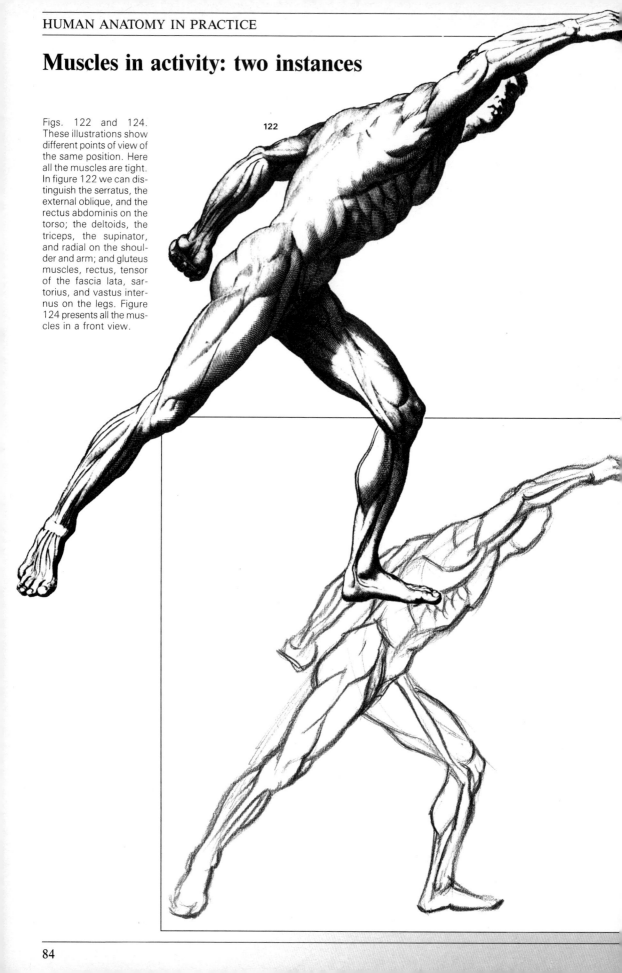

122

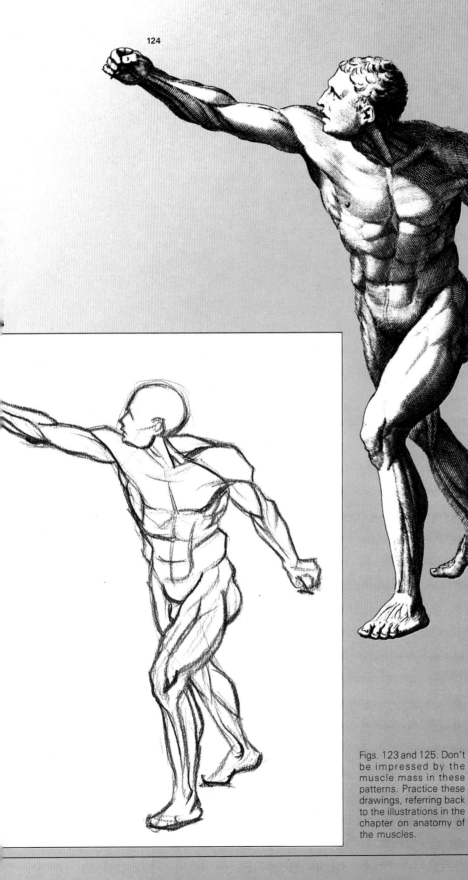

124

Figs. 123 and 125. Don't
be impressed by the
muscle mass in these
patterns. Practice these
drawings, referring back
to the illustrations in the
chapter on anatomy of
the muscles.

The articulated doll: studies

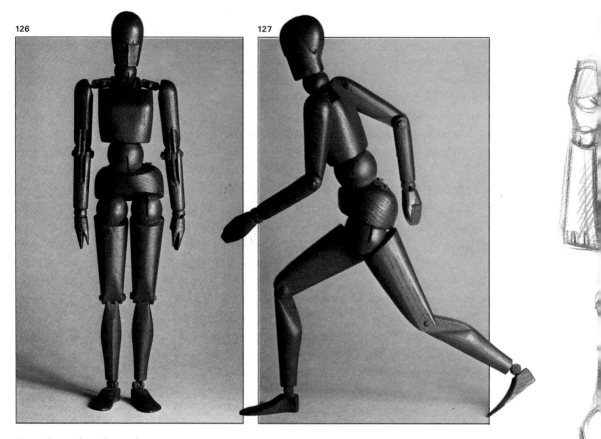

126

127

128

I'm pleased to introduce you to an articulated doll (Fig. 126). Behind its mechanical appearance, there's a loyal friend, always available to pose any way you want for as long as you need. It is eight heads high, the ideal for the human figure (see page 42). The balls and hinges permit the doll to "move" in ways that replicate human movements (Fig. 127). Articulated dolls are commonly used in art schools all over the world because they work—they are an excellent way to study how parts of the body work together in various positions. You will find it useful to have one; check art supply stores for an inexpensive model. Remember, though, the doll shows only relative mass of the head, torso, and limbs. To seriously study figure drawing, you must add what you have learned about anatomy: the bones, muscles, and tendons that give the human body its external appearance.

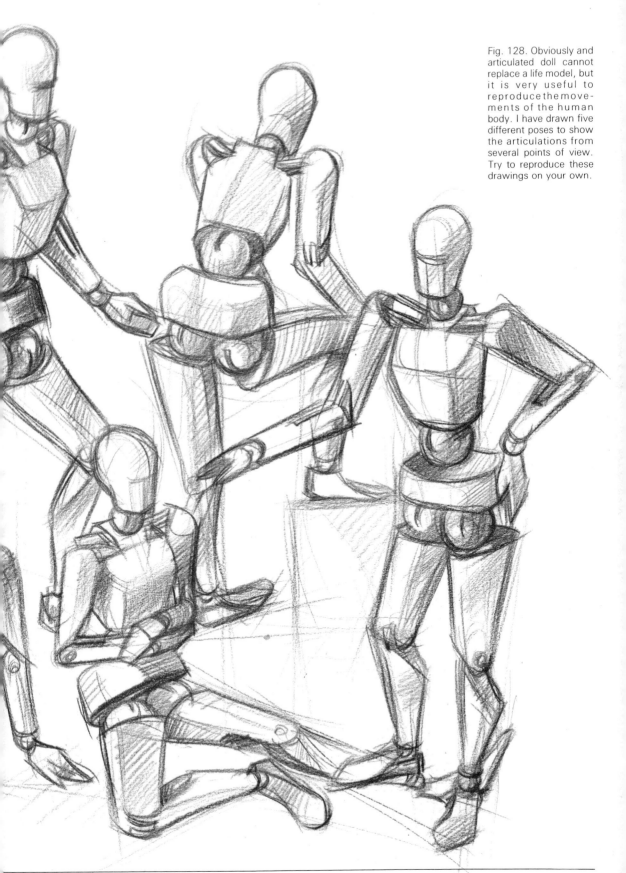

Fig. 128. Obviously and articulated doll cannot replace a life model, but it is very useful to reproduce the movements of the human body. I have drawn five different poses to show the articulations from several points of view. Try to reproduce these drawings on your own.

Practice with the articulated doll: the female figure

129

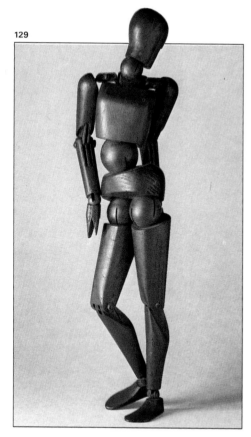

I'm going to draw a female figure using the articulated doll. I know what you're thinking: "You need a lot of imagination!" Well, not so much. Mostly we need a clear knowledge of a woman's proportions and certain basic concepts about anatomy.

Let's start with the position shown in Fig. 129. I chose that pose because it permits us to use most of the doll's articulations, without being too complex or unnatural, and it is not hard to visualize an actual human figure in this pose. Study the relative position of the limbs. Can you recognize that pose? Yes—the ischiatic position: the body weight on one leg, the other leg relaxed and slightly bent. I have positioned the doll in a turning movement, with the shoulders and thorax in a frontal position and the hips in a three-quarter position. The doll's shoulders are parallel, but the more natural attitude would have one shoulder lower, following the inclination of the upper thorax. As you can see, the hip of the stretched leg is raised, and the tho-

rax bends toward that side; this counterposition is typical of the ischiatic position. Now let's start drawing!

First sketch the doll itself (Fig. 130). I used sienna chalks for this preliminary sketch, so that the line here will be muted in the final drawing. In this first phase don't worry too much about getting the shapes just right; the idea is to use the doll to place the various masses properly, since they will be the structure of the drawing.

Now start to fill in the spaces with darker chalk, converting the lines of the doll

Fig. 129 to 135. Use a clear color that allows you to build up the figure and add in further detail. Study the text and illustrations on these pages about the process of a female figure drawing, using as a starting point the sketch of the doll in Fig. 129.

130

more natural lines of a human figure.
n Fig. 131, for instance, bearing in mind
he concepts of anatomy, I am sketch-
ig in the shape of female hips. In the
ame way, following the angle of the
oll's head, I can place the chin, the ear,
he features of the face, and the neck
Fig. 132). Then, to modify the stiff,
nechanical appearance of the sketch, I
ave slightly emphasized the bend in the
ight arm, leaving the hand partially be-
ind the hip (Fig. 133). The doll does not
how the malleoli, so we need to use our
nowledge of anatomy to add them in
orrectly (Fig. 134). This process of "fill-
ig in" continues until the drawing takes
n a realistic look.

n the finished drawing (Fig. 135), no-
ce that some parts that were visible in
he doll are now hidden—the left arm,
or instance. After we add in slight
hadowing to emphasize the volumes,
ho could guess that this was originally
drawing of a lifeless doll?

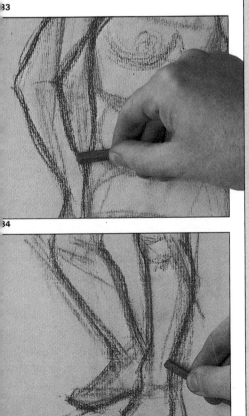

Practice with the articulated doll: the male figure

136

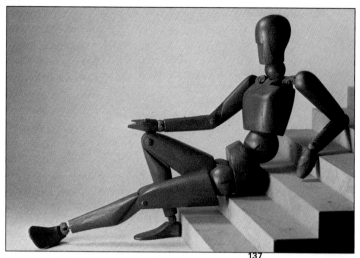

Fig. 137. Sketch the doll accurately. Provide adequate room but not too much blank space. This is a diagonal composition from corner to corner of the box.

Fig. 138. Amend the original doll sketch: bend the head over the thorax. The doll has no details; draw them in, but don't define them too much.

Fig. 139. The figure rests firmly on the arm, so it must present a clear bend.

137

For this exercise of the male figure, we use as our inspiration Michelangelo's "Adam" (see title page of this book). To recreate this famous pose, I positioned the doll sitting on steps (Fig. 136). Now, in this position a figure has several points of support: wrist, elbows, feet, thighs. We will need to take these into account when we fill in the human shape, darkening the areas where the body weight lies. The process is the same as in the previous exercise: first, do a rough sketch of the doll, bringing out the volumes. Since the model stays in a quite horizontal position, it is better to draw with the paper in that direction (Fig. 137).

Then, begin adding the essential features. I improve the doll's pose where it appears too stiff—for instance, bending the head over the chest for a relaxed pose (Fig. 138). The arm must also be corrected (Fig. 139), stressing the folding caused by the fact that it lies on the stairs (the effects of the body weight must be clearly rendered). Also, the right arm (Fig. 140) seems artificial: stretching it just a bit and stressing its volume make its appearance more real.

The figure's weight affects mainly the left leg. The curve of the thigh must show the effect on the spot where it is supported on the stairs (Fig. 141), and the rest of the thigh must be defined by a quite straight line.

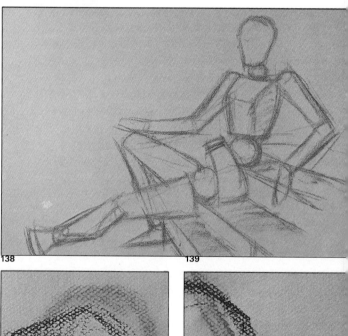

138

139

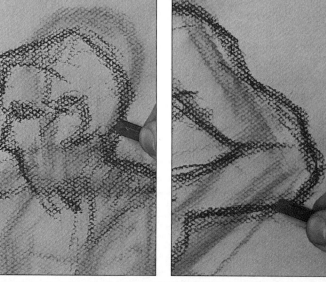

140

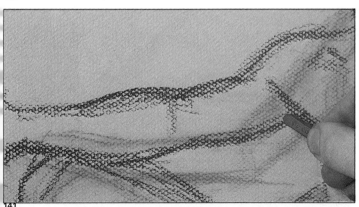

141

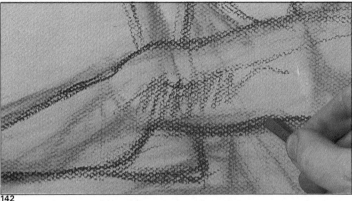

142

Fig. 140. This arm should also be corrected from the original doll sketch. Stretch it out, making the elbow visible. Bring out the different protuberances of the arm and the forearm.

Fig. 141. To emphasize the weight of figure on the thigh, intensify the curve in the area that rests on the stairs.

Fig. 142. Muscular curves and hollows are expressed by shadows as well as lines. Observe that the shadows here are used to define muscle mass and articulation.

Gallery of poses and sketches

Figs. 143 and 146. An easy and attractive technique for figure drawing is sketching in ink on colored paper with highlights of gouache or white chalk. Use this procedure to draw the poses on this page.

Fig. 144. Here we have a human model in the pose of Michelangelo's Adam. If you remember the exercise on page 90, you can redraw this pose without the preliminary step of sketching the doll.

Fig. 145. This figure gives you practice on the shapes of the trunk and arms.

Figs. 147 and 148. These female figures allow you to study the slight muscle mass and the influence of the bone structure on a slim female body. Study particularly the elbows, back, and knees.

Figs. 149 and 150. These two poses put to work most of the muscles of arms and legs. Observe in Fig. 149 the hollow on the back formed by the trapezius and the dorsal, and the structure of the infraspinatus, teres major, deltoids, and triceps on the shoulder.

144

145

143

92

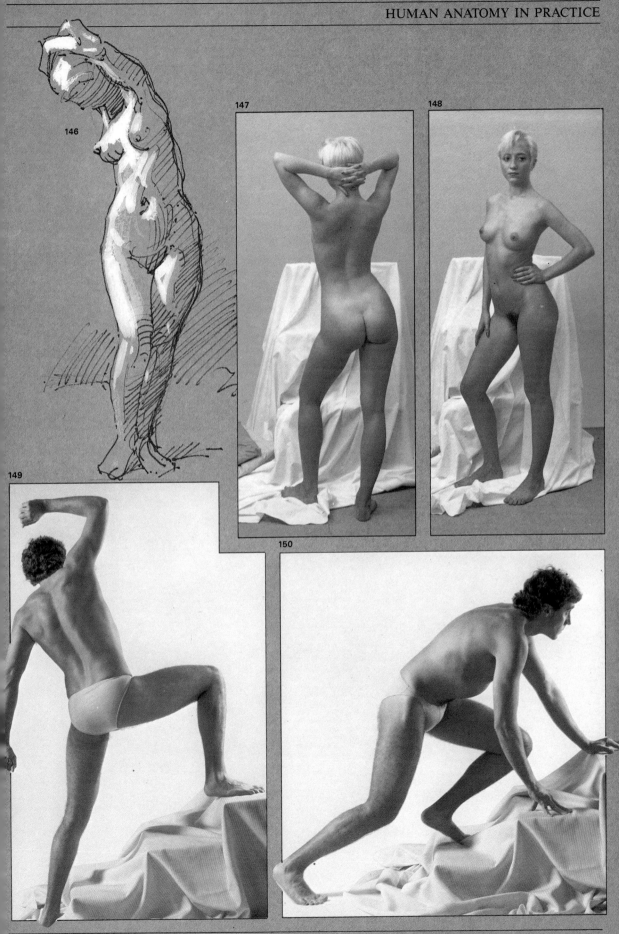

146

147

148

149

150

Studies of the figure: the male model

Fig. 151. These four poses complement the previous exercises on the model. Each presents different muscular tensions, defined by the intensity of the shadow. Remember that the greater the contraction of the muscles, the stronger you should reproduce light and shadow contrast. Study the figure sitting on the right: the pose scarcely implies muscular tension and the shadow does not intensify the muscle contours.

151

Studies of the figure: the female model

Fig. 152. These four drawings show the various masses of female anatomy by means of dark areas that contrast with the light. When you draw these figures, remember that bone structure, rather than muscular structure, must be clear in a female body. Be accurate when drawing the curved limits of the hips and thighs.

152

In his famous "Treatise on Painting," Leonardo da Vinci says: "For man, it's easy to achieve universality, because all the terrestrial animals are similar as for their organs: muscles, nerves, and bones. Anatomy will show that they only differ in width and length." The master refers to using knowledge of human anatomy as a universal pattern, to help understand animal anatomy. Obviously, we won't study the whole animal world, but an anatomic study wouldn't be complete if we completely missed the features of the animals. As Leonardo said, man resembles animals and the structure of their bodies is similar since the internal components are the same.

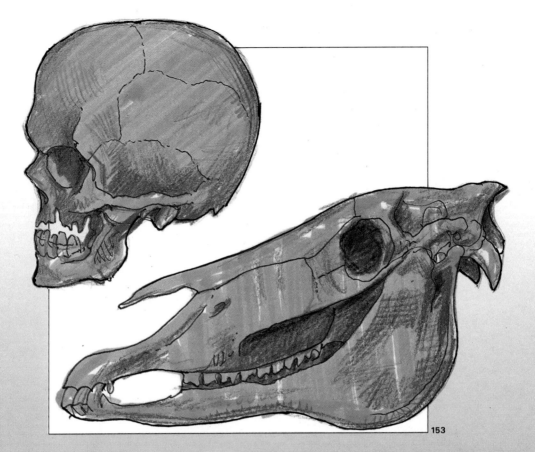

153

COMPARATIVE
—ANATOMY—

Introduction

If we make a simple comparison of the basic factors in the structure of humans and animals, we shall see that the differences are more apparent than real. We find, for instance, that, with the exception of a few entirely aquatic species, all mammals have a head, trunk, and four limbs: in fact, these same features occur in birds and even in some four-footed reptiles.

So, if you know the human form and remember the construction of its skeleton, you will find no difficulty in understanding the forms of most animals. It is simply a question of comparison and study. You must memorize firmly the shape, size, and position of the most important bones, so that when faced with a horse, dog, or bird you can imagine and "see" the internal framework that determines its shape and movements.

154

Fig. 154. To become familiar with the animal structure, try to reproduce these four drawings. Observe that only the most relevant shapes are emphasized; for the moment it is not necessary to stress other details.

Fig. 155. This exercise involves sketching animal figures. The cock has been performed in different techniques: sepia, red chalks, and watercolor. Watercolor provides a colorist quality very suitable for this sort of animal. The heads of the lion and the zebu (the African bull's relative) have also been painted in watercolor: ochre, sienna, and warm colors for the lion; grayish blue and dark reds for the zebu.

The skeleton of mammals

In our study of comparative anatomy, we will compare the skeletons of a man and a horse, bearing in mind that, if we are familiar with a horse's skeleton, we can understand the structures of all quadrupeds, from a cat to an elephant.

Admittedly the heads are different shapes. The horse "face" is proportionately longer than the man's and the cavity within the cranium is smaller: in human terms, you must think of it as having a long nose and a considerable distance between the eyes and mouth. We can see, however, that the two heads are alike in their features and functions. Both are divided into two parts: one part is the cranium and the upper maxillary; the other part is the lower maxillary. Another important detail for artists is that the point where the maxillaries meet lies farther back in animals than in men (and even more so in birds), which enables them to open their mouths much wider.

The horse's neck has a much longer curve than the man's, but both consist of seven cervical vertebrae. Adjoining them lie the scapulae, which are supported by powerful muscles. Like most quadrupeds, the horse has no clavicles. This is why it cannot extend its forefeet sideways as humans can their arms.

As in humans, the horse's cervical vertebrae and scapulae lead to the spine, whose vertebrae can also be seen in the animal's loins and rump. The ribs lead off from these vertebrae to form the thorax or casing for the chest.

Fig. 156. Although both man and horse have seven vertebrae, the horse's are larger and stronger. Overall, the two species have similar skeletons: except for the clavicle, which the horse lacks, the number and distribution of the bones are the same. The size, however, is obviously very different: note the much larger volume of the horse's thorax.

Fig. 157. Study the main differences between a man's head and a horse's head: the smaller cranial cavity of the horse, the greater distance between the eyes and the mouth, and the horse's broader and longer mandible.

156

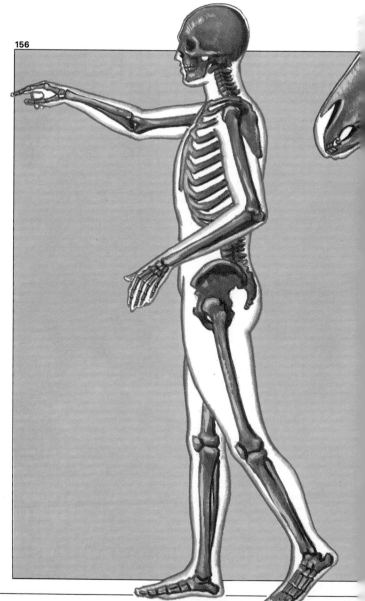

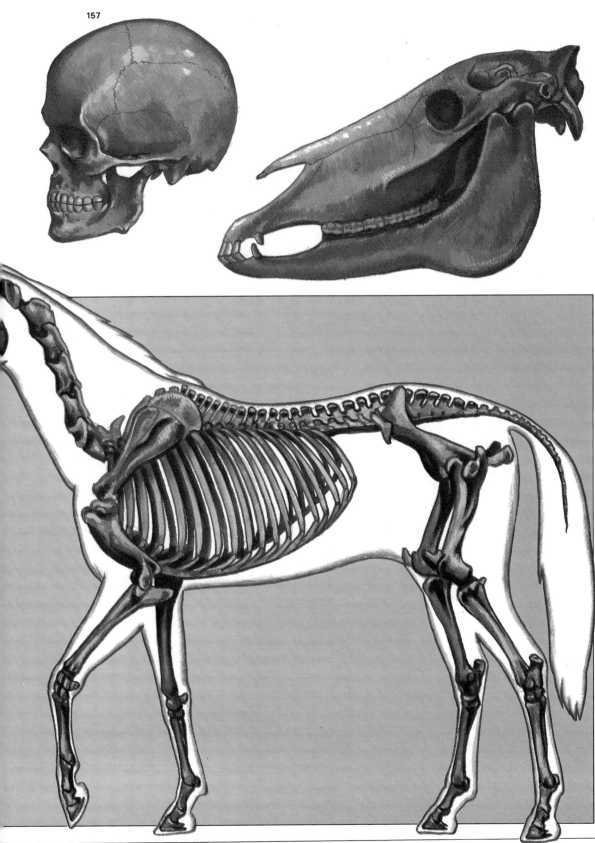

157

The skeleton of mammals

Now let's compare the human arms and legs with the horse's front and hind legs. The special formation and mechanism of these limbs contain the key to understanding the structure of most animals. The horse's humerus is a shorter, stronger bone, which does not appear on the surface since it is encompassed by the mass of muscle in the chest.

The ulna and radius (the human forearm) together produce a similar shape and length.

Below the carpus (our wrist) the horse does not have a wide hand with a palm, but a kind of extended foot (it is really a hand) terminating in four "fingers" welded together to form one: the "nails" of these fingers form the hoof.

Now we can compare the human legs with the horse's hind legs: The horse's femur is shorter and straighter and is joined to the body by masses of muscles. Where the femur joins the tibia and fibula (the latter is almost nonexistent) the horse has a patella that lies (this is important!) almost level with the belly.

The tibia leads to the foot itself, whose heel is as prominent as ours, maybe more so. In the horse the tarsal and metatarsal bones and the toes combine to form one bone or toe terminating in the hoof.

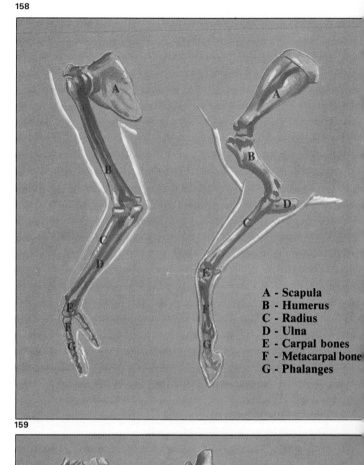

158

A - Scapula
B - Humerus
C - Radius
D - Ulna
E - Carpal bones
F - Metacarpal bone
G - Phalanges

159

A - Pelvis
B - Femur
C - Patella
D - Tibia
E - Os calcis
F - Tarsal bones
G - Metatarsal bones
H - Phalanges

Figs. 158 and 159. The horse's limbs have the same articulations as the man's but their distribution is rather different: the horse's scapula is longer and the bone that resembles the knee is actually the wrist. On the other hand, the animal's patella lies higher than the man's because it has a shorter femur.

Some animals walk like men, in that they place their weight on the sole of the foot; these are called *plantigrades*, and include bears, elephants, and monkeys.

Horses, dogs, cats, rabbits, deer, lions, and similar animals *walk on the points of their toes* and not on their soles, as we do. If you remember that, you will have made a great step forward in understanding the body, shape, and movements of many animals.

Imagine a man bent in almost a kneeling position, like an athlete on his mark. Compare him with the deer shown in Fig. 160. Notice the similarity between the man's legs and the hind legs of the deer; all you need remember then is that an animal's femur is much shorter, while the "sole of the foot" is much longer and thinner.

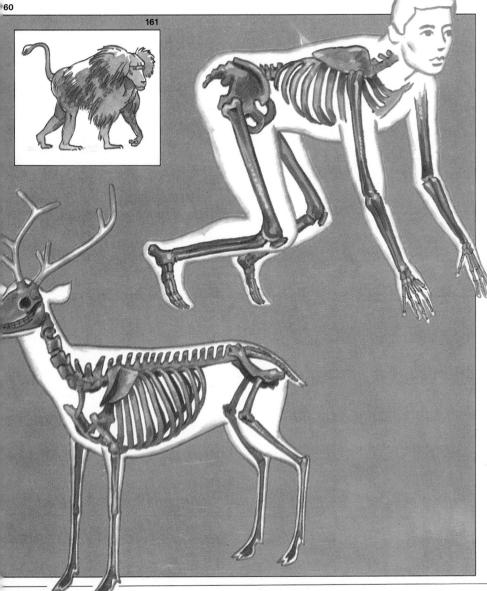

Figs. 160 and 161. If you imagine the skeleton of a quadruped as if it were a man in a kneeling position, you can notice the articulation of its movements. Remember that the most visible articulations in the animal's limbs are the equivalent of man's ankles and wrists.

The skeleton of birds and fish

Study this skeleton of a bird. You will note that, in essence, it consists of a head and dorsal spine, which controls the casing of the chest and the four limbs (the wings are equivalent to human arms). Although the number and type of their bones bear a number of similarities to our own, there's little distance between the casing of the chest and the joints of the lower limbs (Fig. 162A).

The spine occurs in all animals, even those that have only an embryonic thorax, such as fish, whose backbones remind us of ribs; on the other hand, fish lack the complex articulation system that in birds and mammals is equivalent to the limbs.

Fig. 162. The structure of the skeletons of birds and fish accounts for their movements. Reproduce the drawings on this page, bearing this in mind.

162

Fig. 163. As in the illustration in Fig. 155, you can use watercolor for these drawings. Remember that the original sketch must display clearly and easily the animal's shape and mass.

163

The skeleton and muscle structure of a dog

164

We won't deal with the muscular structure of animals in any detail, partly because in many animals the muscles are hidden under the fur or plumage. Besides, by now you know enough about the structure of the body to enable you to draw from life.

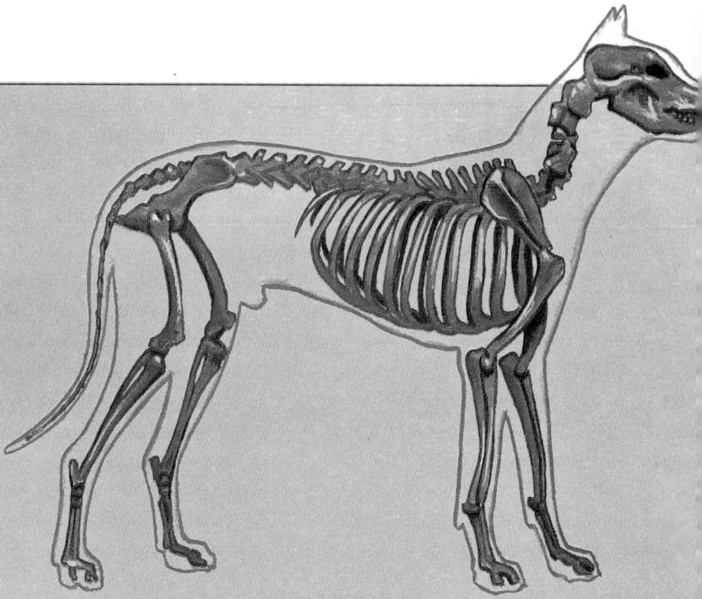

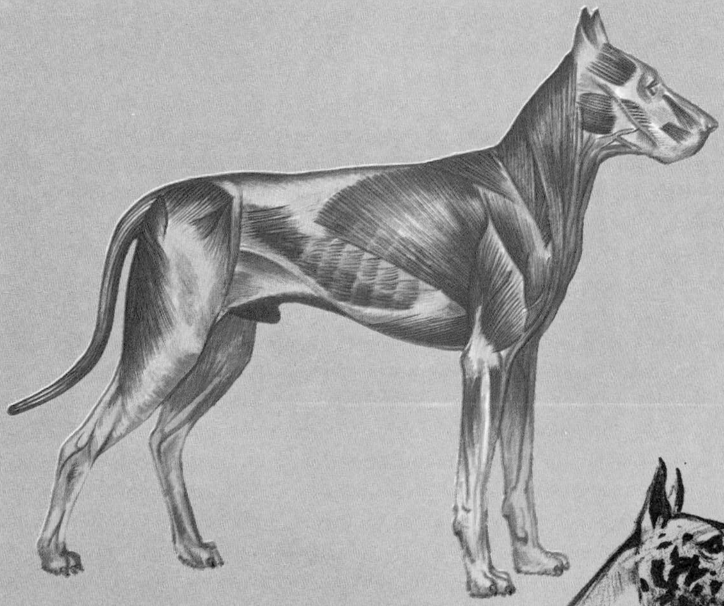

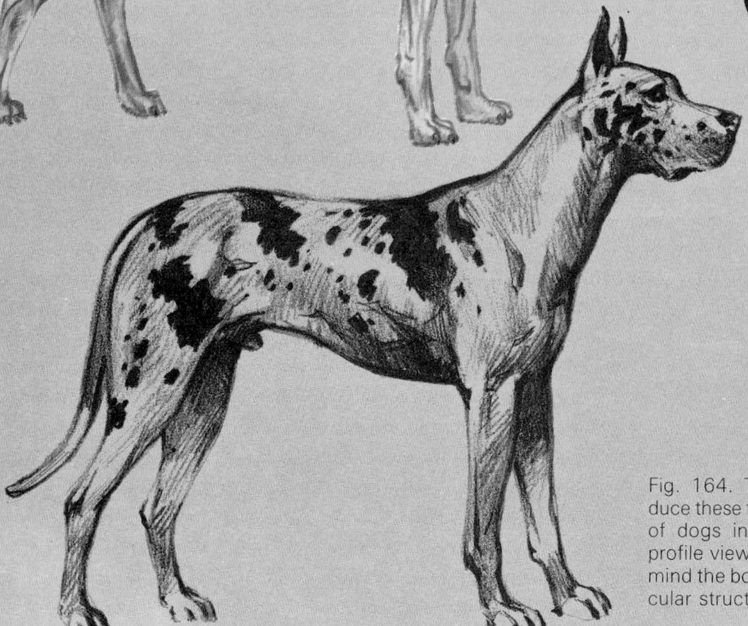

Fig. 164. Try to reproduce these two drawings of dogs in frontal and profile views, bearing in mind the bone and muscular structure.

The skeleton and muscle structure of the cat family

165

The various breeds of dogs have a muscle structure very similar to one another; so do the members of the cat family. Differences of size and fur distinguish a cat from a tiger, a lion from a leopard, but mean no essential change in their muscles. Nevertheless, the development and strength of the muscles are directly related with their size. These animals always present the same characteristics when you draw them.

Fig. 165. The lion's fur hides a muscular structure common to any feline. Draw these two lions, taking it into account. First, reproduce them without the fur, and once you understand the body structure, add the fur to your drawing.

The muscle structure of quadrupeds: the horse

However, some important details about the muscle structure of the quadrupeds (we shall confine ourselves to the horse) should be noted. They have nothing resembling the deltoid muscles, which are the most prominent ones in the human shoulder, and the horse's pectorals are narrower because the function of the front legs is limited—they cannot be extended sideways like our arms. Little can be seen of the muscles on the horse's trunk as compared with humans, who require a more complex and more highly developed and visible system of muscles to maintain their upright position and to bend and twist the waist, torso, and hips. On the other hand, due to the greater tension and power developed as it walks, gallops, or jumps, a horse's legs have a more obvious muscle structure; the os calcis (the human heel) and the *ancon* muscle, used to keep the limb straight like the human triceps, are particularly conspicuous in the legs (Fig. 167).

That's enough. To sum up, we must try to draw animals by remembering that the structure of their bones and muscles is, to varying degrees, similar to the structure of the human. The rest is a matter of practice, of drawing many sketches and studies from life.

Fig. 166. Joaquín Sorolla, *The White Horse*, Sorolla Museum, Madrid. This is one of the most famous paintings by this Spanish master. The magnificent volume of the horse contrast with the lightness of the young boy. Observe the movements of the animal's articulations: you can distinguish the protuberances of the ulna, the carpal bones, and the phalanges in the front limbs, as well as the os calci and the joints of the tarsal and metatarsal bones in the rear limbs.

Fig. 167. Contrary to man's, the muscles of the horse's trunk have minimal contours. Nevertheless, they are noticeable in the upper part of the limbs and the neck of the animal.

166

167

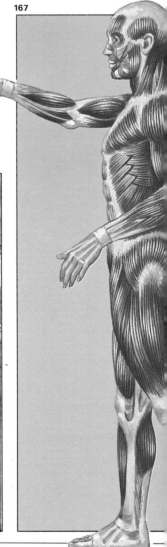

168

Fig. 168. George Stubbs, *Anatomy of the Horse*, British Museum, London. This English artist was the master of sports painting. His interest in horses is similar to the Renaissance interest in the human body. Stubbs drew figures such as this one, where the horse's muscle structure is fully visible, from the dissected corpse of an animal.

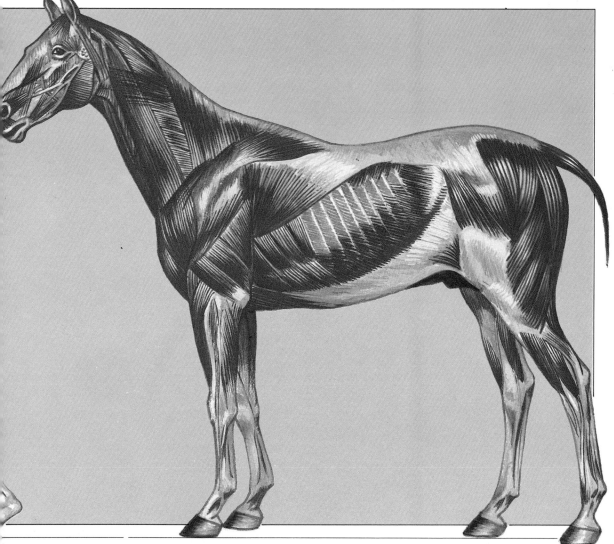

Epilogue

Eventually, the studies you have made, combined with skill at dimensions and construction, will enable you to perform the miracle of drawing a perfect figure. "It's not a miracle at all!" my anatomy professor would say. "All you need is knowledge! Watch me raise my leg. Can you tell me what my dear old sartorius muscle is doing now?" And we could answer—as can you— "Well, it is contracting as far as the iliac crest, pulling up the tibia."

Now it's up to you. If you'll excuse me, my phalanges are weary from so much typing.